Textbook of
MUSHROOM CULTIVATION

The Authors

Prof. Rajni Gupta, She did her B.Sc (M.D. University, Rohtak, 1987), M.Sc (Agra University, Agra, 1989), Ph.D (Agra University, Agra, 1993), Research Associate (CAS-UGC Dept. of Botany Univ. of Delhi, Delhi), Prof. Rajni Gupta joined Delhi University, in the year 1993, in capacity of R.A (CAS -UGC) under Prof. K. G. Mukerji. For five years worked on Mycorrhiza and Mycoherbicides and other fields of fungi. In 2006 she joined Kirori Mal College as a permanent position in Dept. of Botany. She published about 50 Research papers and articles in many invited books and journals published from National and International Publishers like Springer and Plenum. Published about 13 books in the field of Fungi, Microbial Technology, Plant Systematics, Biotechnology and General Life Science, books, Class 10th, 11th and 12th Standard school books under World Bank Project. Presently working on Mushrooms, delivered Key note lectures in FDP,s and National Conference on Microbes. Delivered about 7 talks in various colleges on Cultivation of Mushrooms and How can be Growing of Mushroom can make you Entrepreneur. Worked on Phytoremediation of heavy metals Under Innovation scheme of DU. Presented papers in many National and International Symposia and Conferences. Attended many workshops and presently giving workshops to students and teachers on Cultivation of Mushrooms.

Dr. Ajay Singh, Registrar, Maharana Pratap Horticultural University, Karnal. Regional Director, MHU Regional Mushroom Research,Centre, Murthal. He did his B.Sc, M.Sc and Ph.D from CCS Haryana Agricultural University, Hisar by 1997. Did post graduate diploma in marketing management from IGNOU in 1999. He worked as Scientist- in-Charge, Integrated Mushroom R&D Project, HAIC Agro R&D Centre (a Govt. of Haryana body) for 18 years (2002-May, 2020). PI – AICRP on Mushroom since 2009. PI-NAHEP (IG), MHU Karnal since 2020. Registrar, MHU, Karnal 2018 to date. Regional Director, MHU Regional Mushroom Research Centre, Murthal since July, 2020 to date. Cultivation technology of various Mushrooms, Disease pest management & postharvest technology of Mushrooms. Done excellent research work on mushroom during Masters & Doctoral studies & currently he is Principal Investigator, All India Coordinating Research Project on Mushroom at Murthal & Co-PI of two joint research project on Milky Mushroom & functional foods of mushrooms with IIT, Delhi. Published thirty four research papers in National & International Journals and Eight books on Mushroom & Bee-Keeping. Guided several M.Tech&Ph.D students of Biotechnology of DCR University, Murthal & IIT, Delhi respectively. He has been to abroad thrice for attending International Training Programme & International Conferences.

Textbook of
MUSHROOM CULTIVATION

– Authors –

Rajni Gupta

Ajay Singh

2024

Daya Publishing House®
A Division of
Astral International Pvt. Ltd.
New Delhi – 110 002

ISBN 978-93-5461-719-5

Published by : **Daya Publishing House®**
A Division of
Astral International Pvt. Ltd.
– ISO 9001:2015 Certified Company –
4736/23, Ansari Road, Darya Ganj
New Delhi-110 002
Ph. 011-43549197, 8130496929
e-mail: info@astralint.com
Website: www.astralint.com

Printed at : **Replika Press Pvt Ltd**

Preface

Mushrooms are fleshy-spore bearing fruiting bodies, generally growing above the ground, on land or below the soil also. In ancient Egypt, mushrooms were considered a gift from the storm God set and sent to the earth by lightning. Mushrooms were eaten by Pharaohs. Common people were not allowed to touch them. Pharaohs were so much admirers of mushrooms, that their buildings, jewellery and temple pillars were of the shape of Mushroom.

The ancient Greek, also subscribed to the lightning theory, mushrooms were the plants with invisible seeds. Aristotle classified the fungus as a plant in Natural Philosophy. Mushrooms were considered good to eat but some times due to poisoning, it was risky to consume them. Mushrooms are mentioned in *De-Materia Medica* also. In other civilizations throughout the world *i.e.* Russia, China, Greece, America, Europe mushrooms rituals were practiced. They believed that mushrooms give longevity, super human and strength.

The Chinese were the first to cultivate the tropical and subtropical mushrooms about thousand year back in 600 A D. But real cultivation started in 17th century, French were the first to cultivate button mushroom in caves on horse dung. The first producer of pure spawn were American spawn company of 20th century. By 1914 Button mushrooms was dominating, other mushrooms also came into existence *i.e. Pleurotus, Shiitake, Ganoderma, Auricularia, Cordyceps, Morchella etc.*

In India mushroom cultivation started in 1961, when ICAR funded a scheme on button mushroom cultivation technology at Solan. National Center for Mushroom Research and Training was established in 1983 at the same place under the aegis of ICAR that was later named as National Research Center in 1997. And upgraded to Directorate of Mushroom Research Center (DMRC) in December in 2008. Other

Institutes and Agricultural Universities have under taken R & D activities in button and other mushroom production.

Mushrooms are rich source of nutrients, antiviral, antibacterial, anti cancerous, rich source of protein, low in fat and calories, comes under the category of functional foods. Problem of malnutrition can be improved by the consumption of mushroom.

Employment can be generated by the cultivation of mushroom. Environment pollution problem can be solved by this technology, low cost and high out put technology, getting best out of waste and considered as Sustainable Agriculture .

The present book deals with the information of cultivation, marketing and utilization of earnestly grown edible and medicinal mushrooms, India scenario and world scenario, problem faced by persons. Government policies regarding the cultivation of mushroom has been discussed. This work will be a help for students and researcher. It is need of time to develop start up through cultivation of mushroom and by processing other edible materials by the Youth . Women also can start small scale and large scale industries by growing mushroom and will help in women empowerment.

Authors have tried their best to explain every aspect in detail and make it popular among the Youth.

Rajni Gupta

Ajay Singh

Contents

Chapter 1

Introduction to Mushroom

World of Mushroom is very fascinating, they comes under the class Ascomycetes and Basidiomycetes. Mushrooms varies in their shape, size and colour. They are macrofungi can be seen from naked eyes. In woods they are present wild and can be feast for anyone's eye.

Brain Storming Question

Q. Do you know me, who am I?

Q. Whether I belong to Animalia or Plantae?

Q. Why I am placed in separate clade?

India is an agriculture-based country with cultivated area of about 51.09 per cent. The emergence of food crisis, due to an ever increasing world population and pandemic disease has led scientists and technologists to tap the available sources in finding alternative way of producing cheap and quality food.

Among the various food sources for humans, microbes can be the important source as SCP, and edible fungi can be the largest and important group. Malnutrition is another problem in developing countries can be solved by the consumption of mushrooms.

Mushrooms are edible fungi, have been classified as vegetable but mushrooms are not vegetable it's a fungus. Mushroom due to their flavor, texture, nutritional value and high productivity per unit area have been identified as an excellent food source to combat the malnutrition in developing countries. Reason for quick acceptance of mushroom is its nutritive content. Mushroom can be eaten as a substitutes of meat in taste and flavor. In general edible mushrooms are low in fat and calories, rich in Vitamin B and C, contain more protein than any other food of plant origin and also a good source of mineral nutrients (Bahl, 1998). Currently, high biofuel prices have caused an increase in food prices and food scarcity in many countries (World Bank, 2008).

To improve hunger and malnutrition in a World of rising food process, cultivation of mushrooms is a very reliable and profitable option.

Growing of mushroom is an ecofriendly and part of Sustainable Agriculture as it grows on stubble. In India million tons of agricultural waste is produced each year and burning of that, is the source of air pollution affecting the health of human beings.

Employment can be generated by growing mushroom for the young generation. Small scale and large scale Industries can be developed, women empowerment can be done. In food industry various types of cookies, breads and sauces can be prepared from mushrooms.

As mycelium of Basidiomycetes has a tendency to absorb and degrade heavy metal, it plays a good role in Bioremediation/Mycoremediation and can solve the problem of oil spill in Sea.

Green leather formation, decorative, furniture and packaging material formation, from the mycelium of mushroom is in process. These innovative ideas are very catchy for our young generation.

The artificial cultivation of mushroom technology is a recent innovation, incorporation of non conventional crops in existing agricultural system can help in improving the social as well as economic status of small farmers.

Q. What substrate can be used for growing Mushrooms?

Q. Who coined the term vegan food?

Q. Can mushroom synthesize their own food?

Most of the mushrooms are vegan, since they are not animals or animal product but rather fungi that feed on deadwood and organic matter where they grow. The term vegan was introduced in 1944 by a group of vegetarians Ancient Indian and Eastern Mediterranean societies; Most of the mushrooms comes under vegan food category as they grow on trunk of trees and on agro-waste material.

Structure of Mushroom

Mushrooms are fruiting bodies of fungi, separated from plants and animals group. It can be defined as a macro-fungus which can be epigeous (above ground or hypogeous (below ground) with a distinctive fruiting body. These are large enough, umbrella and spatula shaped, to be seen by naked eyes. They constitute one of the major group of plant kingdom, they do not have chlorophyll pigment, eukaryotic in nature, plant body not differentiated into stem, root and leaves. They can not use atmospheric carbon dioxide for synthesizing their own food. Fruiting body of mushroom is known as Sporocarp or Basidiocarp, divided into Volva, Stipe Pileus and volva (Figure 1.1).

All the edible mushroom, under cultivation falls into two major groups: Ascomycotina and Basidiomycotina.

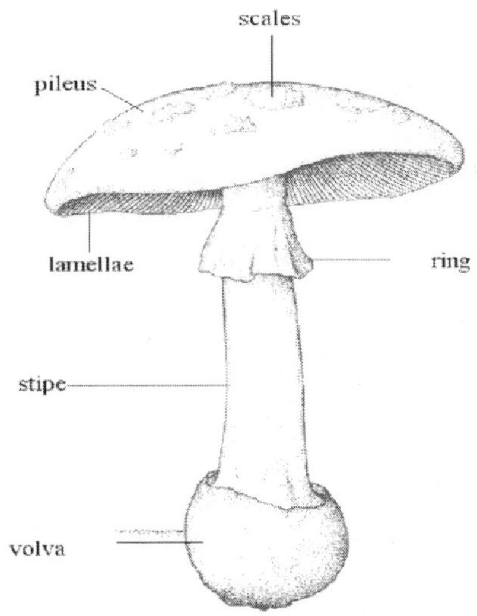

Figure 1.1: Structure of Mature Sporocarp/Basidiocarp of Mushroom.

In the botanical sense, mushroom are fungi which lack chlorophyll and cannot use solar energy to manufacture their own food as do green plants. Mushroom can produce a wide range of enzymes that degrade the complex substrate, on which they grow, they absorb the soluble substrates for their own nutrition. This absorptive nutrition is a characteristic of fungus. Mushroom play a important role in maintaining the life cycle of nature by decomposing and clearing away left over pieces of wood and other organic debris, not easily degradable. Unlike most plants they produce seeds, mushroom produce spores to continue to reproduction process.

In case of mushroom, the spores has no sexual origin (Atkin, 1981). Spores are invisible to naked eye, germinate by producing thread like structure known as mycelium, fruiting body Basidiocarp/sporophore is formed without stem and leaf stage. The fruit body of mushroom is umbrella like or other shapes, sizes and colours, some have additional structures like veil or annulus and a cup or *volva*. In contrast to green plants, most mushrooms are haploid. The diploid phase is normally transient and restricted to the basidium, where meiosis occurs. The meiotic products migrate in the basidiospore of which they bear four basidiospore at the tip. Due to its such a life cycle and saprophytic nutritional habit, mushroom in nature are generally found to grow in places, where dead remains of plants or wooden logs are decaying (Figure 1.2). Majority of mushrooms are saprophytes, there are some which either grow upon living plants as parasite, while other grow in symbiotic association with root parts.

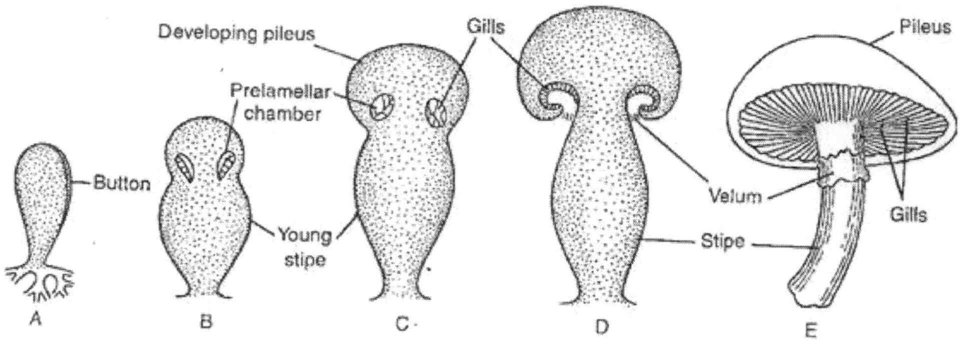

Figure 1.2: Development of Basidiocarp from the Spore.

Q. Explain the structure of Basidiocarp.

Q. How mushrooms are scavenger in nature ?

Q. How they are eco- friendly ?

How Many Species of Mushrooms?

There are large number of fungi are present in the world, estimate is that there are 15 lakh fungi (Figure 1.3). However, mycologists are not been able to study all the fungi and they have studied only 1.1 lakh fungi out of which 14,000 are mushrooms

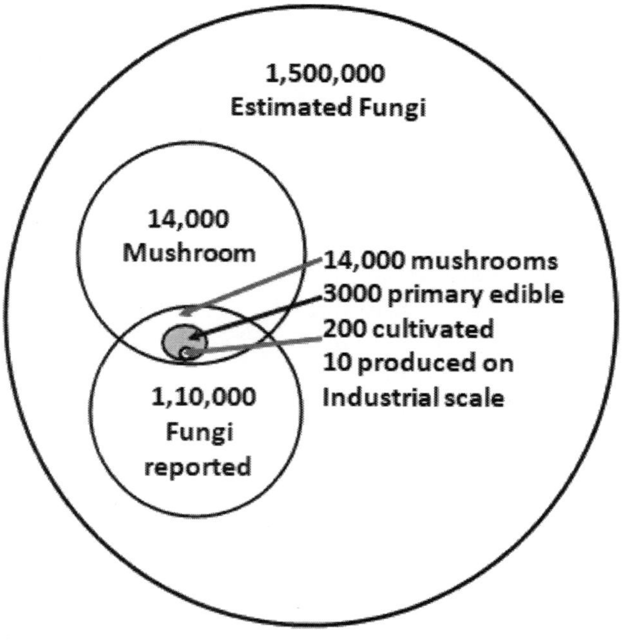

Figure 1.3: Estimated and Actually Reported Fungi and Mushrooms on Earth.

(Actual number of mushrooms may be 10 times of this). All mushrooms are not edible. Some are even poisonous. Out of these only 3000 have been considered to be truly edible. However, it is not possible to cultivate all of these. Despite all the efforts it has been possible to cultivate only about 200 species experimentally, out of which 60-70 are cultivated commercially and about 10 are cultivated on industrial scale.

i. **What is the estimated number of fungi on Earth?**

ii. **How many have been cultivated experimentally?**

iii. **How many species are cultivated on industrial scale?**

The science of mushroom research and development in India, has expanded with passage of time. As a evidence, the yield of mushroom has increased manifold during the last four decades *i.e.* from 2-3 kg/100 kg compost in 1964-65 to 18-20 kg/100 kg compost now. *Agaricus bisporus* popularly known as white button mushroom has the widest acceptability in India and its present production levels is estimated between 1,40,000 to 1,60,000 tonnes per annum. The second most popular is "Dhingri" account approximately 7200-8500 tones per annum. In our country, there is a need to convert the surplus agricultural waste into nutritious food using simple and inexpensive methods.

History of Mushroom

History of mushroom cultivation is about 900 million years old. In Bible it is mentioned as FOOD OF HEAVEN,only Pharoses were supposed to consume mushroom. Even in our Vedas, Somras is mentioned that was obtained from deadly mushroom *Amanita muscaria.* Folkore rarely dies, but it does not change. Indeed many of our most colorfully named mushrooms have been known from few decades.

In ancient Egypt mushrooms were considered a gift from the Storm God set and were sent to earth via lighting. Such treasures clearly conveyed immortality, so could be eaten by the pharoh. Common people were not allowed to touch them. The ancient Greeks also subscribed to the lightning theory, believing that mushrooms were plants with invisible seeds.

Aristotle classified Fungus as a plant in his Natural Philosophy but he did not work on it Without microscopes they would not have been able to see air borne spores and they would have been unaware of the vast net works of mycelia beneath their feet, that give rise to ephemeral fruiting bodies. Mushrooms were considered good to eat, though risky -cases of accidental poisoning were common. Fungus agarikon *Fomitopsis officinalis* was the ancient world medicine that was still recommended by John Gerad in his herbal of 1597 to cure asthma, poor digestion, worms and coughing of blood from the lungus. Nobel were permitted to eat mushrooms, which were cooked in special pots called boletarie. Warriors were also allowed to eat mushrooms that ingested immediately before battle, would improve their strength and endurance. However distressing fairy rings might be folk lore is a pretty much unanimous that is pointers to destroy one. In rich legend to damage a circle by mistake will curse the perperator, kicking of mushrooms invites seven years of bad luck.

According to some reports word Mycology has been derived from Mysena Dynasty. The people of Mycena Dynasty were very much obsessed by mushrooms, their buildings, ornaments, paintings all were of mushroom shapes (Figure 1.4) In some countries like fairy tales amusement parks are of mushroom shape can be seen *i.e.* Netherland is famous for its Mushroom shape amusement park. Domestication of the edible fungi dates back from centuries ago, but the most important development in the field have taken place mainly in the last 80 years. The cultivation accidentally originated sometimes in the middle of seventeenth century in France. French people cultivated mushrooms in caves and limestone pits, revolutionized the method for more scientific cultivation as *Agaricus bisporus* and remained as pioneer exporter of canned mushroom till recently. United Kingdom, United States of America, Japan, Italy and Netherlands picked up its cultivation more scientifically in the recent past. Korea, China, Germany, Belgium, Switzerland, Rumania, Ireland, Bulgaria, Russia and Poland are also aspirants to find their place in the mushroom growing.

In India cultivation of mushroom was started in 60s by the efforts of Dr. S.S. Jain Plant Pathologist working in Shimla on the disease of Apples, Wheat and Potato. The position of farmers was very pathetic there. He wants to work for them so that they can have additional income by growing something. He procured spawn from Japan and France and used the fallen material in the fields like, rotting twigs branches of other fruits, wheat straw and cow dung. After successful growing of button mushroom on the raw material he got published his paper in magazine of

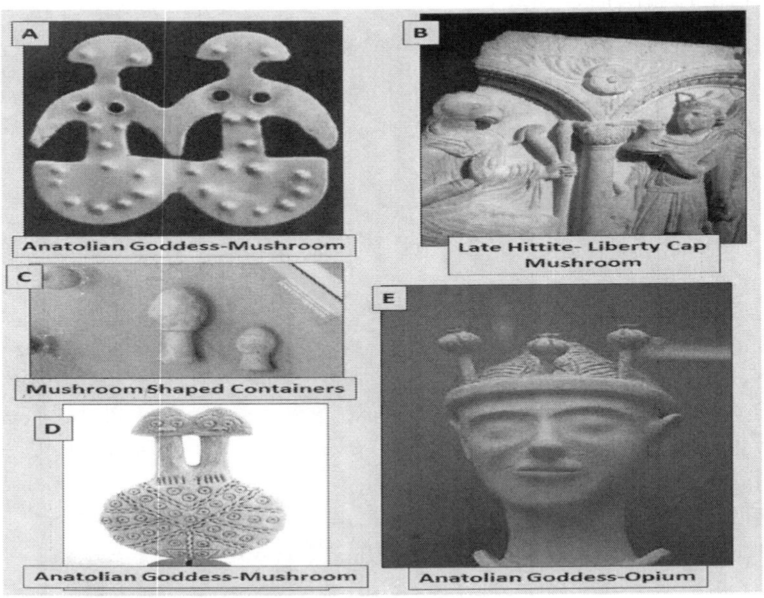

Figure 1.4: Oranaments and Buildings of Mushroom Shape.

HP state Extension Department. After that Dr. S. S. Jain established a Lab in Solan. It became the only important center for training in mushroom cultivation to farmers of HP and other states. Later on many Plant Pathologist became the part of that center in Solan. It was the dream of Dr. S.S. Jain that was started in a room now India is flourishing in this field, doing Million Dollar business and growing various types of mushroom.

At present the mushroom production systems in India is of mixed type *i.e.* small farmers are doing seasonal farming but high-tech industry person are doing farming whole of the year. During seventies and eighties button mushroom was the only crop grown as a seasonal crop in hills. With the development of the technologies for environmental control and increased knowledge of the cropping systems, mushroom production shot up from mere 5000 tonnes in 1990 to over 1,00,000 tonnes in 2010, 1,30,000 ton in 2016 and at present it is likely to be 1,60,000 ton.

Six major mushroom species *Agaricus bisporus, Lentinula eodes, Pleurotus* sp., *Volvariella volvacea, Flammulina velutipes, Auricularia, Calocybe* and *Portebello* (Figures 1.5a-i) accounted for 87 per cent of total world mushroom production. Cellulose, hemicellulose and lignin constitute about 80 per cent of total dry weight of straws. Most edible fungi possess good cellulose and hemicelluolytic and lignolytic enzymes. Proper understanding and identification of such system will help in selection of mushroom strains that have an increased ability to convert most of the substrate into fruiting bodies. Due to continued efforts, country has been able to find its place in the list of important mushroom growing countries. Some common mushrooms grown all over the world are described below.

Q.1 Why we have to grow mushroom?

Q.2 Discuss about commercial use of growing mushroom.

Q.3. How mushrooms are environment friendly?

Q. Mention medicinal properties of mushrooms.

Q. Why mushrooms are recommended for heart patients and diabetic patients?

Health Benefits of Mushroom

Mushrooms fall into some broad categories of some distinct areas. They are of various types.

1. Cultivated mushroom
2. Wild mushroom
3. Psychoactive mushroom
4. Poisonous mushroom
5. Medicinal mushroom
6. Useful mushroom

Scientifically they can be classified as:

☆ *Saprotrophic mushroom*: which grow on dead organic matter i.e., button, oyster, shiitake, reishi *etc.*

- *Mycorrhizal mushroom*: They grow in symbiotic relationship with tree and other crops *i.e.* Truffules, *Porcini* and *Cantharellus*.
- *Parasitic mushroom*: They infect the host and kills them Chaga, Lions mane and Honey fungus.

Mushrooms are the source of extraordinary power used in the preparations of many continental dishes. They have medicinal properties like anti-cholesterol, anti-tumorous, anti-cancerous and anti-diabetic. Mushrooms are the good source of protein, dietary fibres, vitamins and minerals (Khan *et al.,* 1981).

Mushroom contains about 85-95 per cent water, 3 per cent protein, 4 per cent carbohydrates, 0.1 per cent fats, 1 per cent minerals and vitamins (Tewari, 1986).

Mushroom contains appreciable amount of potassium, copper, iron, phosphorus and small amount of Calcium (Anderson and Feller, 1942).

Mushroom protein is intermediate between animals and vegetables (Kurtzuan, 1976). It also contains appreciable amount of niacin, pantothenic and biotin (Subramanian, 1986). Mushrooms are a rich protein source having essential amino-acids and high digestibility. In cereals amount of lysine is very low but it is the most important amino acid in mushroom. Mushrooms have all the nine essential amino acids required by human being. Mushrooms have approximately two times more protein than in vegetables and 4-12 times protein than in fruits. Mushroom can be eaten as fresh raw or cooked in any form unlike other soya products. Considering that number of people suffer from malnutrition, mushrooms can be an important way to combat this problem.

Mushrooms are also recommended for heart patients as they have low fat, no cholesterol, 80 per cent unsaturated fatty acids and some of the mushrooms have compounds (like lovastatin in oyster mushroom) the blood cholesterol can be lowered by that. Mushrooms have low- sodium, high potassium content and folic acid so that it is a suitable food for persons suffering from high blood pressure and Anemic disease. As Sodium content is very low so its good for the ailments of heart and kidney diseases.

The mushrooms are also considered delight of diabetics as these are a low calorie food, which have number of anti-oxidants and no starch. These are also rich in fibres, good for the intestine and digestive system. These are also a very good source of vitamins especially vitamin B complex. Mushrooms are the only vegetarian source of vitamin D. If they are exposed to UV light after harvest just for 15 minutes or expose to sunlight, the content of vitamin D increases more than 100 times. Mushrooms have high amount of riboflavin, thiamin and cyanocobalamin (Vitamin B12), that is usually found in animals. Vitamin – niacin and anti- pellagra is equivalent to found in pork or beef, that is considered to be the richest source of vitamin.

Many bioactive compounds have been isolated from the mushrooms. As *Ganoderma* is very famous since ancient times in the folklore medicines from

China. Many bioactive compounds have been isolated from the fruiting bodies, spores and mycelium of the *Ganoderma*. The bioactive compounds include steroids, fatty acids, nucleotides, proteins, triterpenoids, polysaccharides, sterols and *etc.* These compounds are antiviral, antibacterial antifungal, anti-cancerous, antiulcer, antiaging, immunity development and helps in sleep promotion. Ganopoly was the first ingredient that was isolated from the *Ganoderma*, that is a polysaccharide alpha or beta (1-3) (1-6) Glucans and heteropolysaccharides with various combinations of glucose, mannose, galactose, xylose, fructose and arabinose. FIP, a small molecule protein isolated from *Ganoderma lucidum* that play a great role as antitimour, anti-allergy, stimulation activity on immune cells to produce a variety of cytokines. Still *Ganoderma* is widely used as valuable herbal medicine and in health supplements. Herbal tea and herbal coffee is available in market in the name of Reishi, known for its high medicinal value and nutritive value. Cordycepin is the compound obtained from *Cordyceps militaris* is very nutritive helps in the control of diabetics and anti-cancerous in nature. Many labs in India are working on it. Pleurotin obtained from *Pleurotus ostreatus* is an anti-inflammatory and with anti- oxidant activity compound. Nebrodeolysin is hemolytic and antiviral in nature. Eryngin and pleurostrin are anti-fungal and anti-bacterial in nature. Laccase enzymes present in oyster mushroom is antiviral and ribonuclease present in it, inhibits breast cancer and hepatic cancer.

Mushroom can be grown on agricultural waste and industrial waste. Waste can be recycled into food and environment may be less endangered by pollution (Hayes, 1978).

In flood affected areas where soil is not rich in nutrient, crop cannot be grown, there cultivation of mushroom is a good alternate.

Mushroom cultivation is a short duration crop, vertical cropping can be done and land can be saved, much labour is not required, cost effective so it is welcomed by poor farmers. At present mushroom production has increased around the world.

Mushroom Cultivated on Industrial Scale

Agaricus bisporus

It is a predominately crop of temperate region of the world. It is mostly confined to Europe, Asia, North and South America, Africa and Australia. Today the world production of cultivated mushroom reaches almost 4 billion tons a year and grown in almost 90 countries. For that, Pilot stations has been created along with scientific research institutes and consultation centers,where research efforts are carried out and mushroom cultivation specialists are trained with this consultation help is given to mushroom industries. In Russia, there are some specialized farms for mushrooms cultivation which produce about 7 thousand tons of mushroom in a year. In our country the mushroom production is making its steps. Every year more private small farms for mushroom cultivation are created, laboratories that work on mycelium production are developing (Figure 1.5a).

Figure 1.5a: ***Agaricus*** **(Button Mushroom)**

Lentinula edodes

This mushroom is commonly known as shiitake in Japan. Due to its distinct flavour it has enjoyed a good International reputation as a food. It has taken an important place in the agricultural industry of Japan, Korea and China, but other countries still depend for their supply from forests (Figure 1.5b).

Figure 1.5b: ***Lentinula*** **(Shiitake)**

Pleurotus Species

This mushroom is commonly known as Oyster mushroom. It is a lignocellulolytic fungus and grows naturally in the temperate and tropical forest on dead and decaying logs. They also grows on drying trunks of deciduous or coniferous woods. With more than hundred per cent biological efficiency coupled with its pleasant flavour and excellent drying and keeping qualities, its a good source of food. Commercially available species are *P. ostreatus, P. sajor-caju, P. florida, P. sapidus, P. eryngii, P. cornucopiae, P. flabellatus, P. pulmonarius* and *P. eous*. This mushroom is domesticated in France, Germany, Hungary, United States, Taiwan, Japan, Korea, Thailand, Italy and India. *Pleurotus* is available in many colours,adaptive to variable temperature and can grow on wide variety of agricultural waste. As waste can be converted in to food it helps in over coming the problem of environmental pollution (Figures 1.5c-d).

Figure 1.5c: *Pleurotus florida* (White Oyster)

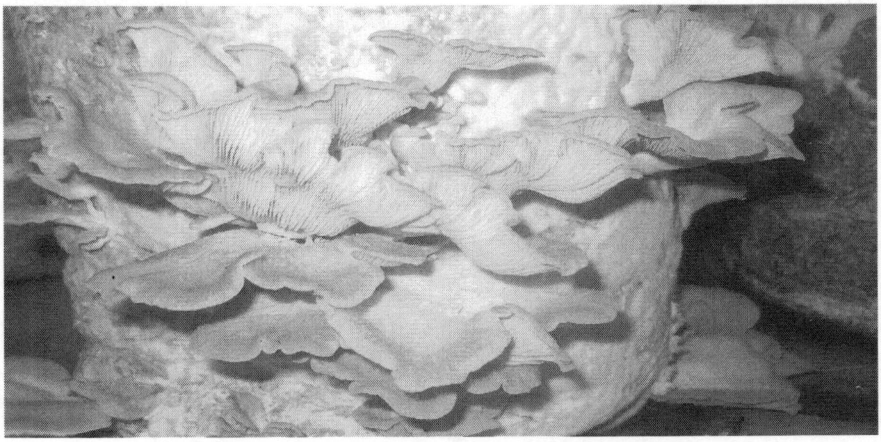

Figure 1.5d: *Pleurotus djamor* (Pink Oyster)

Figure 1.5e: *Auricularia* (Ear Jelly mushroom)

Figure 1.5f: *Calocybe* (Milky Mushroom)

Figure 1.5g: *Portebello* (Brown Mushroom)

Figure 1.5h: ***Volvariella***

Figure 1.5i: ***P. columbinius*** **(Grey Oyster)**

Volvariella Species

Volvariella volvaceae is an edible mushroom of tropics and subtropics. It was first cultivated in China and later on introduced into South-East Asian countries. It is a fast growing mushroom with simple technology. Cultivation of this mushroom can be developed by cottage industry in rural and sub-urban areas of tropical and subtropical regions of the world (Figure 1.5h).

Auricularia Species

It is a highly prized mushroom in Chinese market. It was primarily grown in China and other Asian countries, specially in Japan and Philippines. Due to its delicacy and rich nutrient, both species of *A. polytricha* and *A. auricula* are very popular in dried form. The optimum temperature for its growth is 10-36° C but can grow between 20-30°C, with 85 per cent humidity (Figure 1.5e).

Tremella fuciformis

This species are commonly known as Jelly fungi or silver ear fungi. Generally it is cultivated on the wooden logs of *Mangifera indica, Acacia confusa* and *Morus alba*. It is commonly grown in Asian countries. In the dried form it is exported to Japan and America. *Tremella* have high medicinal value due to its capacity in curing the ailments like tuberculosis and blood hypertension. The optimum temperature for mycelial growth in 25°C and during basidiocarp formation favourable temperature in 20-28°C with 90 per cent humidity.

Flammulina velutipes

This mushroom is commonly known as Enoki mushroom or winter mushroom. According to world production it ranks fifth in number and generally grows in temperate regions. It is one of the most tastiest mushroom. It can be sold as fresh and in dried form.

Calocybe indica

This mushroom commonly known as Milky mushroom and a new addition to mushroom growers. It was first reported from West Bengal. This mushroom species is suitable for hot and humid environment and can be cultivated indoor in high temperature and high humidity area. It grows well at a temperature range of 26-38°C and 80 per cent relative humidity. Fruiting bodies can be harvested after 34 days. Advantage of this mushroom over other mushroom is easy method of cultivation, less investment, very attractive fruiting body, pleasing milky white colour and long shelf life (Figure 1.5f).

Morchella Commonly known as Gucchi, delicacy of Kashmir. That belongs to Ascomycota, not cultivated in India, but trying to cultivate it. Generally it is found in woods of Kashmir, H.P. and Uttarakhand. Many varieties of *Morchella* are available in India.

Q. Mention the names of some commercially grown mushrooms.

Q. Mention the botanical name of dhingri mushroom.

Q. Mention the botanical name of milky mushroom.

Q. What are the conditions required for the growth of mushroom.

Market Scenario

Marketing of mushroom in India not yet well organised. Fresh mushrooms have very short shelf life, that's why they can not be transported to long distance without proper refrigerated transport facility. It can be sold at local market and around the production area. The cultivation of white button mushroom under the controlled condition can be grown through out year. But during the months of Dec.-Feb, it experiences market problem that time about 75 per cent of the annual production comes in market and are forced to sell their produce at highly low prices. If producer sells it directly to retailer it have its own limitations. In India consumption of mushroom is very less in comparison to other countries. In India it is less than 50 gm per person, than one kg in various countries. Serious efforts has not been made, to promote the product to strengthen and expand the market in order to increase its consumption.

Now a days, mushroom is going to be in demand for processed and fast foods. Mushroom can be canned to meet the damage in the off-season and in the non-producing areas. Regarding the problems of sale/export of canned mushrooms, serious thought has to be given to bring down the cost of production of mushrooms and its processing in order to compete in the international market.

Some of the suggestions to solve the marketing problems of mushrooms especially of white button mushrooms in India are given below:

1. Expand the Market Area and Strengthen the Demand
   a) Popularize mushrooms using ICT as delicacy with nutritive and medicinal value, on mass media like Doordarshan also ads and posters.
   b) Break consumer resistance by creating awareness in new areas. Demonstration of receipies and free samples in new areas. Distribution of free recipie booklets.
2. From Co-operatives for Sale
   - ☆ Create cold storage facility
   - ☆ Create refrigerated transport facility
   - ☆ Create processing facility
   - ☆ Create distributor function for big cities.
3. Good prepacks for eye appeal.
4. Decrease the cost of production and bring down the sale price to boost the demand.
5. Train retailers about handlings, storage, food value and recipes.
6. Approach supermarket, chain vegetable stores, Mother dairy retail counters for retail sale.

7. States should fix minimum support price.
8. Public sector market, processing and export organisations should come forward.
9. Assured supply throughout the year at a reasonable constant price is key to good marketing. Efforts must be made to diversify and cultivate different mushrooms throughout the year along with cultivating some of the important mushroom during off-season under controlled condition.
10. In a village or co-operative the crops should be time-scheduled to get a daily reasonably uniform production to avoid glut on a day, this is required to meet the demand. The marketing system has to viewed as a value chain whereas components are taken care of. The increased production should get translated into economic given for mushroom produces. Awareness about consumption and health benefits of numbers of mushrooms available for cultivation is limited, their demands is also less. Hence multipronged strategy is required *i.e.* Expand market, increase demand and organised marketing and form cooperatives (Wakchaukre, 2011).

The National Committee on the use of Plastics in Agriculture (NCPA-1982) has recommended green house technology for adoption in various regions of India. The indoor cultivation techniques and green house technology, inspired villagers to not only set up mushroom units but also diversity. While many farmers have started growing carnations side by side, other have turned to vegetables cultivation. This technique will not only provide a gainful employment to the rural youth and women but the cost of production per unit area will greatly reduced, the mushroom production in India will increase and will help in recycling the agrowaste (Jagdeesh *et al.,* 2018).

QUESTIONS

1. **Explain the structure of *Agaricus*.**
2. **How many types of edible mushrooms are available in market?**
3. **Discuss the History of mushroom cultivation.**
4. **What strategy we can follow to promote mushroom cultivation?**
5. **Discuss about the nutrient value of mushroom.**
6. **Why to grow mushrooms?**
7. **Mention the bio-active compounds present in the mushrooms.**

Chapter 2

Facilities Required for the Cultivation of Mushroom

Mushrooms are of various types, it can be saprophytic, parasitic, edible, non edible, wild and mycorrhizal in association. They can be a primary decomposers or secondary decomposers depends upon the type of composting method of mushroom grown in lab or in mushroom farm. For the cultivation of mushroom, the mushroom house walls and roofs must be properly clean, free from spider webs and small holes should not be there in wall. Wall must be made of proper concrete so that there can be no gaps in the walls, proper sanitation and cleaning of that area can be done. Post harvest room and packaging rooms should be properly cleaned. For the preparation of spawn proper lab is required with autoclave, laminar air flow and incubator.

1. Mushroom Spawn Preparation

Mushroom spores grown on grains commonly known as "Spawn" in the mushroom industry. It is formed by the expansion of mycelium on the substrate that can be grains of wheat, paddy, sorghum, corn,rye and coffee beans. To achieve better productivity of mushroom, good quality of spawn is the first pre-requisite condition. Small growers and farmers generally purchases it from the spawn laboratories, but big commercial units make their own spawn. As the farmer takes care in selecting the variety of seed and check quality of seed before sowing of the crop, in the same way precaution to be taken during purchasing and transporting of mushroom spawn by mushroom grower. If anyone is familiar with the sterile techniques or having specialized knowledge of culturing of mushroom, can prepare their own mushroom culture and it will help in reducing the cost.

Steps in Spawn Production

1. Before the spawn production, one has to grow mycelium on the medium under sterile condition from either spore or tissue of mushroom. Potato Dextrose or Czapek-Dox's Agar medium can be used for the growth of mycelium. Medium should be free from fungal, bacterial and other microbes.
2. For preparation of spawn, laminar air flow or clean air room can be used. Spawn can be produced in closed and sterile lab and it can not be produced in open room. A hygienic, sterile enclosed air space is required for the preparation of spawn.
    - ☆ As mushroom spawn is not easily available every where and it is expansive also. Grower must know about the good quality of spawn, long distance transport of spawn is not easy to do. Proper care is required during transportation, as temperature matters during the transportation, to avoid deterioration, refrigerated transport in preferred.
    - ☆ For the preparation of mother culture of mushroom proper, healthy and good quality of mushroom is to be taken. After surface sterilization with 0.1 per cent Sodium hypochlorite small part of mushroom is placed on the medium in the Petriplate. Medium in the Petriplate can be PDA/CDA Malt Extract Agar. Petriplates are properly sealed after inoculation, covered with aluminium foil and placed in Incubator at 22±3°C temperature. After 4-5 days, mycelium growth can be seen in the Petriplate. That young mycelium can be used for the preparation of spawn and further can be sub cultured in test tubes with medium to preserve or to be used later on.
    - ☆ Seeds for the spawn production can be used *i.e.*, wheat, rice, sorghum, corn, ragi, coffee bean, saw dust. Grain to be used in spawn production should be of good quality and free from any infection. Generally wheat grains are used as a substrate for this purpose in North India. The wheat grains are boiled partly in water till becomes lightly soft, not up to bursting. After this water is decanted through the sieve and parboiled grains are allowed to dry in shade by spreading on sterilized surface.
    - ☆ After evaporation of surface water from the grains, 2 per cent gypsum and 0.5 per cent Calcium carbonate are added and mixed thoroughly and filled in the polypropylene bags/glass bottles.
    - ☆ PolyPropylene bags/glass bottles are plugged using non-absorbent cotton plugs. Poly fill can be used as a plug in place of non absorbent bottle.
    - ☆ PolyPropylene bags/bottles are sterilised in the autoclave. Autoclave works like a pressure cooker. They are autoclaved at 15±lb pressure

for two hours. Temperature and pressure are maintained in the autoclave. The temperature inside the autoclave must be 121.5°C during sterilization.

- After completion of autoclaving, cooling of the autoclave contents are taken out and left for cooling the bags/bottles. After cooling of the bags and glass bottles, the mycelium of mushroom is inoculated in the grains under aseptic conditions inside the laminar air flow/or in sterile room.
- After this, incubation part takes place. Bags/Bottles are incubated at 25±°C for 2-3 weeks for the colonization of mycelium on the grains. After 3 weeks spawn becomes ready in the bottles/bags.
- Further, this mother culture/spawn can be used in labs for the culturing as an inoculum. Same spawn should not be used after 2-generation as an inoculum; that deteriorates the quality of mushroom. Polypropylene bags are generally used for the commercial scale spawn production, as it is easy to handle, transport and lowers the cost. During growth, bags can be shaked gently twice or thrice to avoid lump formation.

Precautions to be taken during Spawn Production

1. Area to be used for the inoculation must be sterile, free from germs.
2. Sterile your hands properly by 70 per cent alcohol.
3. UV light must be kept on of, laminar flow prior start of work. But it should be switched off 15 minutes before one starts the work.
   - Every equipment like forceps, marker, scissors, cello tape should be properly sterilized.
   - Surrounding area must be clean, dust free and sterilized.
   - Mother spawn should not be used after 2-3 generations. That will affect the yield and quality.
   - Spawn should not be more than one month old.
   - Bags and bottles should be properly labeled after inoculation, with date, name of the species inoculated.
   - Last not the least grain should not be of very small size, otherwise mycelium will be of sticky type.

Spawn can be available in various forms such as it can be developed on grains, saw dust, can be in plug form or in liquid form (Figure 2.1).

The life cycle of mushroom starts from spores but for spawn formation, mycelium is preferred rather than spores due to variations and mutations. Producers of spawn must be qualified because it is the basic and more important step of any crop. It should be properly maintained and sub-cultured. New strains can be developed with development in genetic variations methods and techniques.

Figure 2.1: Culture of Mushrooms (Spawn).

Know yourself

Q.1 What is spawn?

Q.2 Define mother culture?

Q3. Which type of grains can be used for the formation of spawn?

Q4. Why small seeds not be used for spawn formation?

Q5. Which type of bags are used for the spawn formation?

Q6. In which form spawn can be available?

Q7. What is the incubation temperature during spawn formation?

Long Questions

Q1. What precaution to be taken while transporting spawn?

Q2. What can be the constituent of CDA, PDA and Malt Extract Agar medium?

2. Culture Media

Microbes they are ubiquitous in nature, can be seen everywhere. They can be present in soil, air we breathe, acquatic and terrestrial in nature. To grow these microbes in lab, proper nutrients and environmental conditions are required. The nutrients required for the growth of microorganisms can be micro or macronutrient, known as culture media.

On the basis of their consistency it can be of various types.

1. Solid media
2. Liquid media/broth
3. Semi solid media.

Solid Media

In solid media solidifying agent is Agar-Agar, that is a complex carbohydrate mainly a galactose. Agar serves as an excellent solidifying agent and very supportive in nature for the growth of fungi, but for bacteira it is not so supportive.

Agar plates and Agar slants can be made for the isolation and maintenance of fungal mycelium (Figure 2.2).

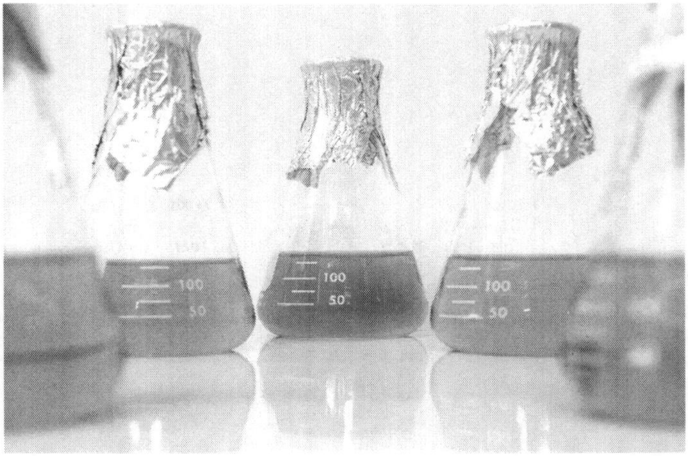

Figure 2.2: Solid Media

Potato Dextrose Agar, Czapek's Dox Agar and Malt agar medium comes under solid medium.

Liquid Media/Broth

In this type of medium other constituents are present but solidifying agent Agar is absent. That is known as liquid media/broth. Broth medium is used to grow mycelium on the liquid medium, organisms can be separate in the mat form.

PDB and CDB can be formed by this method.

Semi Solid Media

This type of medium contains very less amount of Agar-Agar. It can be about 1 per cent so that it can be of semisolid form.

On the basis of composition of media it can be of synthetic, semisynthetic and natural culture media.

Synthetic Media

It is also known as chemical defined media, this is prepared by purified and particular composition of ingredients.

Semisynthetic Media

In this type of medium particular composition of nutrients with natural content takes place. Malt Extract Agar and Potato Dextrose Agar medium comes under this category.

This type of medium contains vegetable juices, blood, peptones *etc.*

Enriched Medium

Addition of extra nutrients in the form of eggyolk, blood, serum to the basal medium makes them enriched media.

Selective Media

It is used for selective or for particular microbes, addition of dyes, chemicals, change in pH, antibiotics makes it selective in nature.

CDA Medium

Sucrose – 30 gm

Sodium Nitrate – 2.00 gm

Di potassium phosphate – 1.00gm

Magnesium sulphate – 0.50gm

Potassium chloride – 0.50gm

Ferrous sulphate – 0.01gm

Agar-Agar – 18gm

D.W. – 1000 ml.

PDA

Potato	–	200gm
Dextrose	–	20gm
Agar-Agar	–	18gm
DW	–	1000ml.

During the preparation of broth, Agar-Agar is not added.

Brain Storming

Q.1. Define media.

Q.2. How many type of media can be used in culturing of fungi?

Q.3. What is the source of nitrogen in medium?

Basic Instruments Required in Lab

Weighing Balance

Weighing balance is the basic requirement of any lab to weigh the chemicals in small quantity and large quantity. Weighing balance can be of crude type and sensitive electronic balance (Figure 2.3).

Figure 2.3: Weighing Balance

Simple weighing balance with single pan can be used to weigh large amount of material upto 500 gm– 1kg can be easily done.

On the electronic sensitive weighing machine can be used for small quantities of chemicals such as in milligram or gram.

Electronic Analytical balance can be used to weigh small quantities of chemicals and samples precisely and quickly.

It can be single pan or double pan type.

pH Meter

pH meter is an instrument used in the measurement of H- ion concentration of any aqueous solution. By which the acidity and alkalinity of the solution is determined and maintained (Figure 2.4). In a pH meter single or multiple glass electrodes are connected to a bulb, which is selective to hydrogen ions and further attached to a metal rod. An electrochemical potential is generated, when the bulb with electrodes are immersed into the solution. Hydrogen ions present in the solution exchange with positive charge on the electrode. This electrochemical potential generated on electrode is displayed in terms of pH units on digital display.

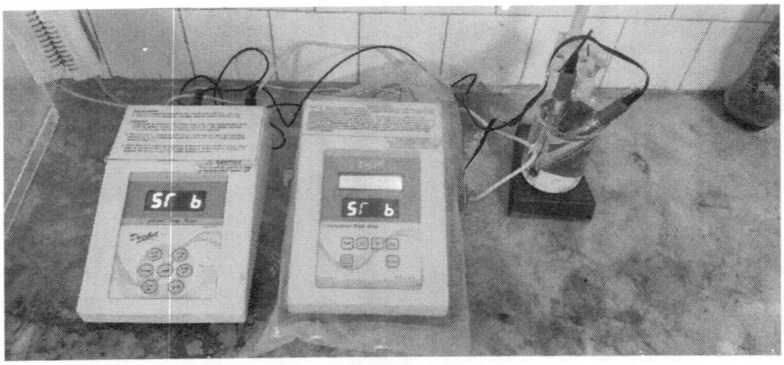

Figure 2.4: EC and pH Meter

Uses

pH meter is used for the measurement of alkalinity, acidity of buffer solutions, soil and culture medium *etc.*

Laminar Air Flow

It is a closed chamber used for the transfer of sterilized material and inoculate fungi and bacteria. Dust that float in air is the main source of contamination so many types of microbes can be present in that. Laminar flow is fitted with HEPA (High efficiency particulate air filter) so as to make it germ/dust free (Figure 2.5).

Figure 2.5: Lamina Air Flow

Principle

An sterile working environment is created by the filtered air in the laminar flow. UV lamp and HEPA filters help in the sterilization of working area. Two types of hood can be there *i.e.*, Horizontal and vertical hood. In the horizontal hood filter pads are present at the back of working area, while in vertical hood filter pads are present on the top of the cabinet. HEPA filter pads present in the laminar flow hood entraps microbes and dust particles from entering in to work station so that chamber can not get contaminated with the microbes present in the outside air. An UV lamp present in the chamber can be used to sterlize the chamber before use. The platform made up of stainless steel can be sterilized by 70 per cent alcohol prior to use. Chamber can be kept closed from the front side during sterilization UV light should of be switched off 15 minutes prior to use to avoid any side effects.

Uses

Laminar flow is used to conduct the processes like plating, transfer into culture tube of microbes fungal/bacterial spawn inoculation, tissue culture and genetic transformation experiments.

Autoclave

It is used for the sterilization of liquid substances media, chemicals, surface sterilization of glass wares. Autoclave is the center point of any lab. It works like a domestic pressure cooker by three factors *i.e.*, pressure, time and steam (Figure 2.6 a&B).

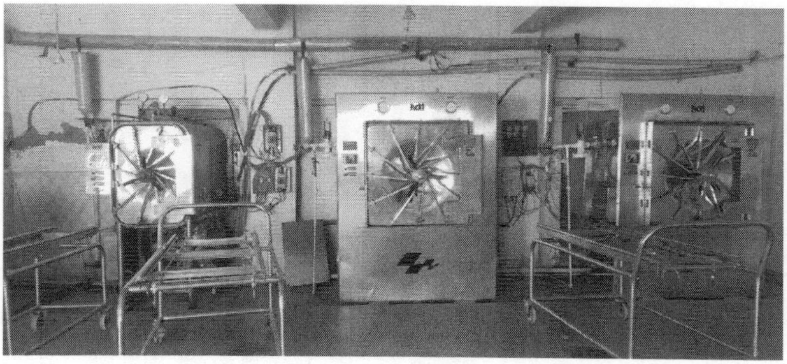

Figure 2.6a: Horizontal Autoclaves

Figure 2.6b: Vertical Autoclave.

Autoclave can be of Various Type

1. Portable externally heated autoclave that can be heated on gas.
2. Vertical autoclave.
3. Horizontal Autoclave.

Principle

It works on the principle of moist heat sterilization process. For the sterilization of chemicals and glass ware water is added in the autoclave. Heating element present inside the Autoclave makes the water boil and high pressure of ±15 lb is maintained in it. High pressure inside the autoclave ensures rapid penetration of heat into deeper parts of the materials to be sterilized.

The moisture present in the steam causes accumulation of proteins of the microbial contaminants irreversible loss of their activities. Time duration of sterilization depends on the type of material being sterilized. For ex. glasswares, liquid media is sterilized for 15 minutes at 15±lb pressure and 121.5°C temperature while soil should be sterilized for 60 minutes.

While using the vertical Autoclave water should be in sufficient amount heating element must be dipped in water otherwise bottom of the autoclave will get dried and further heating will damage the autoclave. Water level inside the autoclave must be maintained. All the locks in the lid must be properly tied to reduce the leakage of heat.

Uses

Autoclave is generally used to prepare the media, sterilize the glass wares and disinfect the used plates with the discarded medium.

Hot Plate

This plate is made up of iron which have a heating element on the lower side. Regulator is present in it by which the required amount of heating can be maintained.

These days sensitive or touch hot plates are also available (Figure 2.7).

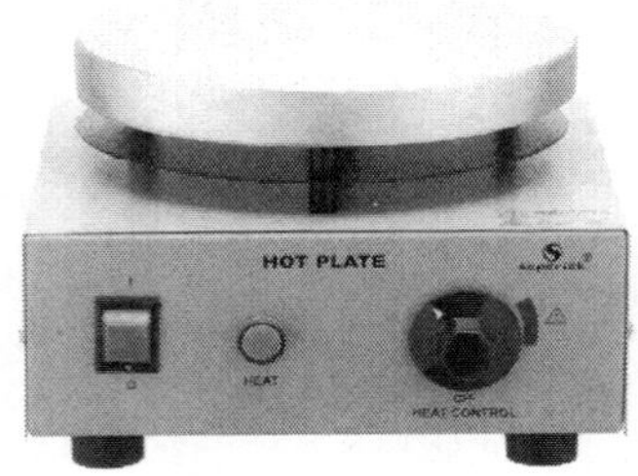

Figure 2.7: Hot plate

Principle

It works on the Principle of production of heat using flow of electricity. Electricity run through the coil, which have high level of electrical resistance. Through this resistance in the coils converts electrical energy to heat energy, due to that coils releases heat.

Use

Hot plate can be used to heat or melt the chemicals and reagents. It can be used to avoid spill of chemical, that can happen in water bath.

Incubator

It is an instrument used in microbiology laboratories for the growth and maintenance of microorganisms and cultures. Particular/optimal temperature, moisture, pressure and aeration can be provided to the microorganisms as everything is in controlled conditions (Figure 2.8).

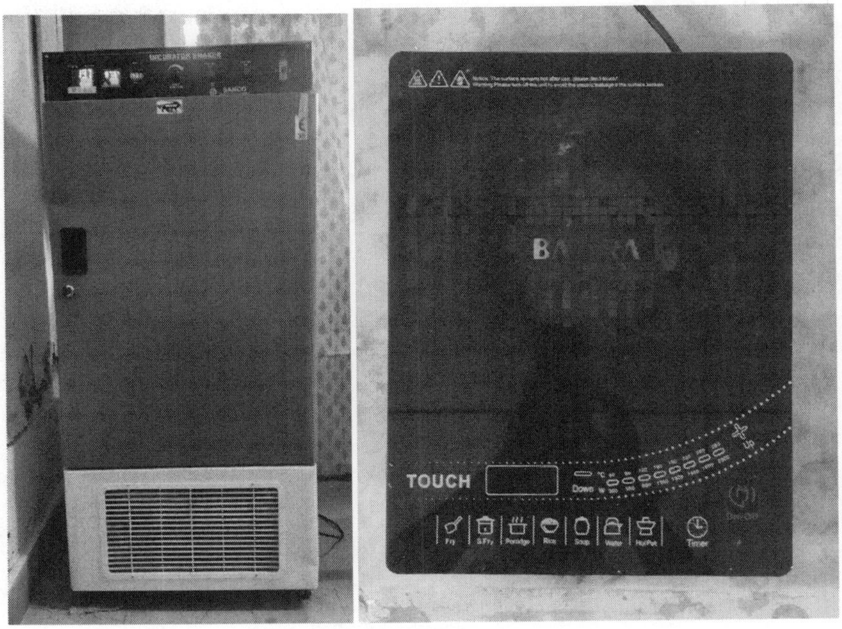

Figure 2.8: Incubator

Principle

Under controlled conditions, condusive atmosphere for the cultivation of microorganisms can be maintained. Heating element and fan is present inside the incubator which maintains the optimized temperature for the growth of microorganisms.

Control panel at the top of the incubator helps in maintaining the functioning of the incubator.

Glass door helps in the clear visualization of the inside of the incubator during incubation without change in the temperature. Double door locking system helps in tightly close the cabinet. Gasket present inside the incubator keeps the incubator air tight between the cabinet and door. On the digital control panel temperature indicator, time, power indicator and heating indicator, on and off switch helps in controlling and maintaining the temperature for the growth of microorganisms.

Uses

Good quality Incubator is the strength of any lab. Used for the proper growth of fungal and bacterial culture, spawn formation and mycelium running on the compost.

Most of the modern BOD Incubators are programmable, that do not need temperature setting by trial and error method. Incubator maintains automatically according to requirements.

Deep Freezer/Refrigerator

Refrigerators are generally used to store chemicals, reagents, solutions, antibiotics and material like mushrooms or spawn at low temperature or even at zero degree centrigrade cultures and sub cultures also can be preserved in it.

Water Sprayers

Maintaining of moisture content of compost and casing soil during fruiting body formation of mushroom is primary requirement, as mushrooms are very sensitive to temperature and moisture. Economic value of mushroom depends upon it's quality and size, so proper water supply is required for that purpose. To grow 1 kg of mushroom 2 liters of water is required, 1 liter of water evaporates. Most of the water is obtained from the compost, casing soil, and from the moisture present in the room. 80-90 per cent humidity is maintained in the room by the sprayers. Various types of sprayers are available in the market. Mist of water vapour can be maintained by fixing the water supply unit with fine nozzles on the roof or on the stands of mushrooms. On small scale, knapsack sprayer, foot sprayer can be used which can spray water,one bag at a time (Figure 2.9).

Figure 2.9: Water Sprayers

Compost Tunnel

It is a fully automatic system in which temperature and fresh air supply can be maintained.

Tunnel is insulated with panels and air duct for the pasteurization of compost.

Composting Air Handling Unit

This unit helps in providing appropriate temperature to compost inside the tunnel so that bacteria can start the natural biological process inside the tunnel and compost formation takes place and that is of good quality.

As control panels are present in AHU so temperature can be controlled automatically inside the unit (Figure 2.10).

Figure 2.10: Air Handling Unit

Straw Pasteurization Machine

It is an automatic, power saving, hot water pasteurization machine for pasteurisation of substrate for growing of edible fungi.

Pasteurization capacity of the machine is 50 kg to 100 kg straw pasteurization per cycle. It can be done by the recycling of hot water.

Significance

Rate of infection can be avoided by spawning and filling of bag with machine. Loading of straw can be done easily every process is automatic, so it is time saving.

Helps in cost cutting but reducing the number of labours. 2-3 persons are sufficient for handling the unit. As water recycling system is present in the machine it reduces the water wastage and consumption of energy.

Compost Mixing and Bag Filling

This machine can mix different components of compost. Filling of bags of required weight and size with compost can be done easily. Machine is made up of stainless steel (Figure 2.11)

Figure 2.11: Mixing of Compost & filling of bags

- ☆ Impellers present in the machine can mix the different components of the compost properly and the compost formed is of fine quality.
- ☆ It reduces the labour cost.
- ☆ It is time savier as filling of bags to be done by semiautomatic machine.
- ☆ On the requirement basis, weighing and filling of bags can be done easily.
- ☆ Excess moisture can be removed, as machine is auto cut at 40°C.
- ☆ Increases the shelf life of mushroom.
- ☆ Clean and clear system, one can avoid washing of mushroom to increase the shelf life.

Proper moisture level is also required for the growth of microbes/bacteria for composting, that can be done with the help of sprayer or can be done manually.

Functions

- ☆ Proper mixing of compost can be done easily and routinely with higher mixing rate per hour.
- ☆ By maintaining proper aeration and moisture level bacterial growth can be maintained composting can be done properly.

Compost Turner

Proper mixing of compost and its components can be done easily by the compost turner. It provides proper aeration and maintain moisture level of compost (Figure 2.12).

Figure 2.12: Compost Turner

Various compost turner are:- Tractor based, Motor based and Self Propelled Engine based.

- ☆ Good quality of compost depends upon the mixing of components properly. For large scale production of compost machines are required that can not be done manually by the labour. Large JCB machines can be used for the proper mixing of components for the purpose of growing button mushrooms.
- ☆ For the good digestion of cellulose component of straw, aerobic bacterial growth is required further in turn proper aeration is required that can be done with the help of blowers.

For large scale growing of mushrooms such as button mushroom, oyster and milky mushroom various types of farm machineries are available. So whole of the year good quality mushroom can be grown.

Know yourself

Q.1. How bag filling can be done on small scale and large scale?

Q.2. How compost can be maintained at large scale?

Composting Unit Machinery

(i) Compost turner
(ii) Compost mixing and bag filling
(iii) Straw pasteurization machine
(iv) Compost tunnel
(v) Composting AHU (Air Handling Unit)
(vi) Spawn dozing and Bag filling

(vii) Water spraying machine.

(viii) Environmental control system

Postharvest Equipment

(i) Mushroom Dryer

(ii) Vacuum Refrigerator

Chaff Cutter

It is a machine used for cutting the Wheat and Paddy straw into small pieces of 3-4 cm in length known as chaff cutter. Since ancient times, this type of machine is used for cutting the fodder and it is hand operated. Wooden handle placed on side used to move the cutter, which is very sharp and made up of cast iron. Knife is of curved shape, tray is placed below the knife so that cut material pass through there and fall in front of that in the mound form. Length of the straw can be adjusted by fast movement and slow movement of the knife. Generally it was used in olden days, these days many companies are manufacturing automatic, semiautomatic and computerized machines (Figure 2.13).

Figure 2.13: Chaff Cutter

Boiler

A boiler is required for the supply of steam to the pasteurization tunnel and canning unit. Capacity of boiler depends upon the process to be used whether for pasteurization or for canning. For canning purpose additional capacity of boiler is required in which external accessories to be fitted like chimney, tanks and pipes (Figure 2.14).

Figure 2.14: Boiler

Bioreactor

It is a large fermentor tank or container, in which microorganisms are encouraged to grow or multiply on large scale. By the development of modern biotechnology. Various types of bioreactors are used for the production of various bioactive metabolites. Bioreactor can be of suspension system, immobilization system, air lift reactors or stirred tank reactors (STR). STRS are the most preferred and conventional bioreactors in research. It is made up of a tank, mixer and series of devices for the measurement and control of different parameters such as pH, temperature, dissolved oxygen concentration, mixing and supplying of all the nutrients (Figure 2.15).

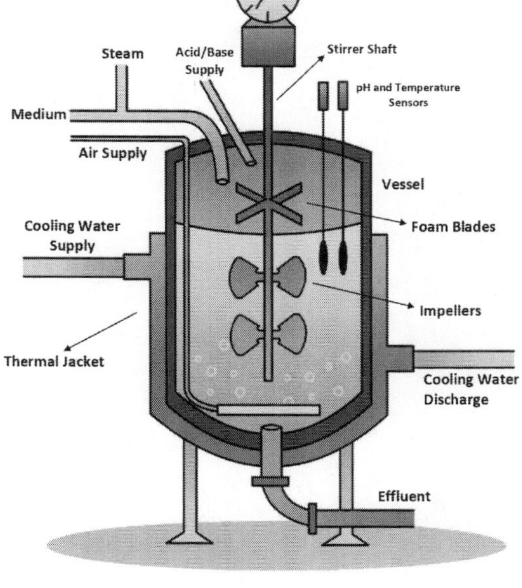

Figure 2.15: Bioreactor

Knackpack Sprayer

Knackpack sprayer is fitted with a large tank, pump and large air chamber. This type of sprayer can be hanged on the shoulder. Functioning of pump can be done by pumping from one hand and spraying from other. By keeping the pump in operation, pressure can be maintained uniformly. By using this sprayer about half hectare area can be watered in a day (Figure 2.16).

Figure 2.16: Knackpack Sprayer

Foot Sprayer

Foot sprayer also works on the same principle of knackpack sprayer but in this case pressure inside the pump is generated by foot pressing and releasing the pedal mounted on the sprayer (Figure 2.17).

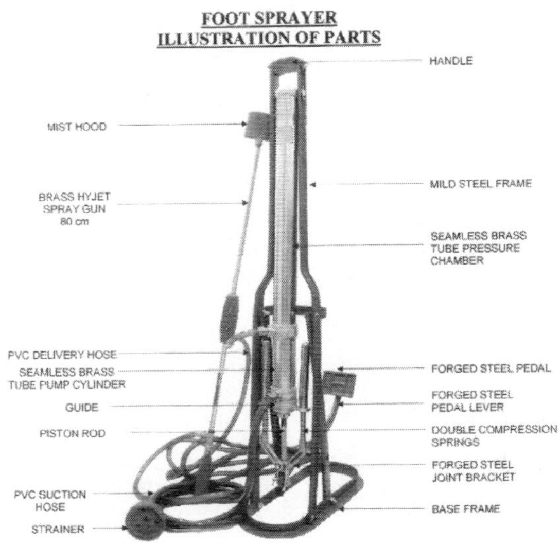

Figure 2.17: Foot Sprayer

Other sprayers can spray many rows of mushrooms beds at the same time with the help of nozzles. Nozzles can be fitted at the top or side of the stand and can be connected to the main type through the gate value. With the help of knobs height of nozzles can be changed. These days smart sprayers are available in the market, which sprays water through soft mist, in ten minutes more than 100 bags can be watered.

Test Yourself

Q.1. How bioreactor is a boon for large scale industries?

Q.2. What type of facilities or machines can be used for large scale units?

Q.3. How humidity of mushroom house can be maintained?

Q.4. Explain about various types of sprayers.

Mushroom Processing Unit

Due to high perishable nature of mushrooms. Processing of the mushrooms is required to increase their shelf life. Drying and canning is the main post harvest technique for which various machines are required.

Dryer

Drying is a process in which the moisture from the mushrooms is removed so that it can be protected from microorganism and spoilage of mushrooms can be avoided. Various types of Dryers are available: (Figure 2.18).

Figure 2.18: Heat Dryer of mushroom

Freeze Drying

During freeze drying removal of water takes place by direct sublimation from the frozen state to the vapour state. It is a three step process (Figure 2.19).

Figure 2.19: Freeze Dryer of Mushroom

1. Ice formation takes place of the water present in the mushroom.
2. By sublimation these ice crystals are removed.
3. Little amount of water present in the mushroom can be removed by evaporation through freeze dryer.

Low pressure is maintained in the chamber and heat radiation is supplied in trays. The temperature of the mushroom is not raised above 0°C and the sublimated water is separated from the mushroom (Sahay and singh, 1994). During this process size, shape, flavour and nutrients of the mushroom are retained.

Cabinet Air Dryer

In an enclosed structure, trays are placed. Mushroom is placed on the trays, and heated air passed through them. In this system trays are placed one above the other. For uniform heating, fans are there it can be vertical or horizontal in position to spread the heat in the cabinet. Temperature is maintained at 60°C for 8 hrs (Figure 2.20).

Disadvantage of this system is that drying does not uniform, colour and flavour of mushroom is not retained.

Figure 2.20: Cabinet Air Dryer

Fluidized Bed-Dryer

It is a good quality drier, in which uniform heating of mushroom takes place. Generally it is used for the drying of oyster mushrooms for 7 hours, at 50°C by this drying of mushroom takes place. A high velocity drying air, fluidized the material. Due to which suspension formation takes place and high amount of moisture is separated out (Figure 2.21).

Figure 2.21: Fluidized Bed-Dryer

Solar Dryer

Sun energy can be used for drying of the mushroom. In direct sunlight mushrooms can be dried. But solar drier's are also available, in which solar energy can be used for drying of the mushroom efficiently. For the proper ventilation, fans are there so that browning of the mushrooms can be prevented (Figure 2.22).

Figure 2.22: Solar Dryer

Honey Comb Resin

This type of dryer works on the Principle of Honey Comb Technology. These are 25 per cent energy saver, control over dew point and temperature. It have Bry wheel desiccant rotor, crafted for very low dew point up to (–) 60°C. Moisture laden air enters into the unit though inlet and moisture is absorbed by the Brywheel. Moisture is removed by the physical adsorption process. Through the process outlet dry, dehumidified air is delivered to the conditioned area. Moisture is removed by the movement of hot air blown through the reactivation segment of the rotor.

Mushroom have to be stored in dry cool places and humidity control must be maintained while drying, packaging storage and transport. If humidity will be uncontrolled then mold growth can be seen on mushroom, RH level must be below 30 per cent, otherwise black mould will cover the mushrooms.

Canning

For long term storage of mushrooms canning can also be another process. Microorganisms responsible for spoilage and food poisoning can be killed or maintained by this method. By putting mushrooms in a container with hot water is known as canning. Button mushroom can be used in sliced form or as a whole for this processing. *Cantharellus* as soft and delicate they can be properly washed before canning. Some precautions can be taken during canning – lid of the Jar must be proper, not too tight not too loose. Air space should not be there, otherwise material can be spoiled. Smell of mushroom must be proper, if stored material smell is not good that can be avoided or discarded

Natural texture and colour of the mushroom must be maintained. Colour of liquid present in the jar must be clear if any change in colour that is the sign of spoilage, can be discarded. Date of manufacture, rate, quantity and date of expiry must be mentioned properly on the bottle and Jar. It can be at the top, bottom or on the side of bottle or jar.

Freezing

Freezing of mushrooms can be another method of preservation. Make small slices of mushroom after proper washing. Saute them with a pinch of oil and butter. Cool the mushroom and pack it in plastic containers and freezer bags.

Mushrooms can be boiled for 10 minutes, cool and pack into the containers with some liquid and seal the container.

Cantharellus and morels can be preserved in this way easily.

Powder Form

Mushroom can be preserved in the powder form. Wash the mushroom properly, slice them. Dry the mushroom in a pan so that water can evaporate, then place them in the oven by spreading them in a single layer, heat at 200 degree centigrade in the oven. When it becomes dry completely, crush it or grind it in mixer to make powder and preserve it in air tight container this can be used in soups and sauces.

QUESTIONS

Q.1. Discuss about short term storage of mushrooms.

Q.2. What do you understand by Mushroom processing unit ?

Q.3. How drying of mushrooms can be done? Explain various types of drying methods of mushrooms.

Q.4. Explain how canning method is useful in preserving the mushrooms.

Q.5. Can powder form of mushroom will help in increasing the business of mushroom?

Chapter 3

Cultivation of Button Mushroom (*Agaricus* sp.)

Classification

Kingdom	–	Eumycota
Phylum	–	Basidiomycota
Class	–	Agaricomycetes
Family	–	Agaricaceae
Order	–	Agaricales
Genus	–	*Agaricus*

Agaricus and other fungi described as mushroom, that produces stalked structures with a cap known as club fungi.

Introduction

Mushrooms belongs to the unique group of fungi endowed with the ability to biodegrade lignocellulosic waste to produce fruiting bodies which are palatable. They are of various shape, size, texture and structure. Culturing of mushroom falls into the thrust area of research "conversion of waste into food". Mushroom production represents the examples of solid state fermentation, with the advantage of separating of the biomass most easily from the growth substrate. They have capacity to degrade cellulose, hemicelluloses and lignin. Mushroom is a table delicacy all over the world, more than 2000 species are edible. Among that some are grown commercially while others are grown at industrial scale. Button mushroom, *Agaricus bisporus* contributes more than 37.2 per cent production and consumption. In Northern India button mushroom in grown on industrial scale and constitutes 95.0 per cent of about total production in this part of India. It is a large fruiting body, devoid of chlorophyll can not make their own food but they are saprophytic in nature. Button mushroom comes under the category of secondary decomposers and absorb its nutrient from the lignolytic substrate on which it grows *i.e.*, straw of wheat, paddy, sorghum, paper, coffee bean compost.

Test Yourself

1. How growing of Mushroom is sustainable in nature?
2. On which substratres button mushroom can grow?
3. Why button mushroom is secondary decomposers?
4. Are button mushroom autotrophic in nature ?

Structure and Life Cycle of Button Mushroom

Fruiting body of button mushroom is umbrella shape and white in colour. Long stalk like structure in known as stipe at the base of which thin thread like structure is present known as Rhizomorph or mycelium. Upper fleshy part, which is rounded in form is edible part-known as pileus. Below the pileus., gills are present in superficial view gills looks like a comb. During detailed studies of that various layers are present hymenium, sub hymenium, and trama. On the basidium basidiospores are present which are formed by the meiotic divison so spore is n in number; Fruiting body is known as Basidiospore or Basidiocarp (Fig. 3.1).

Figure 3.1: ***Agaricus*** **(Button Mushroom)**

Life Cycle

The vegetative muycelium is composed of many interseptate hyphae. Reproductive phase is initiated by the formation of small knob like structure at different points of interwoven mycelial strands. The knob like structure increase in size and divides into two part. From the lower part, mycelium formation taken place and in the upper part- gill primordia formation takes place and further increase in size, veil like structure ruptures and forms the umbrella shaped fruiting body mycelium during the fruiting formation is 2n in number, it can be homokaryotic or heterokaryotic, it depends upon the type of mycelium (Fig. 3.2) .

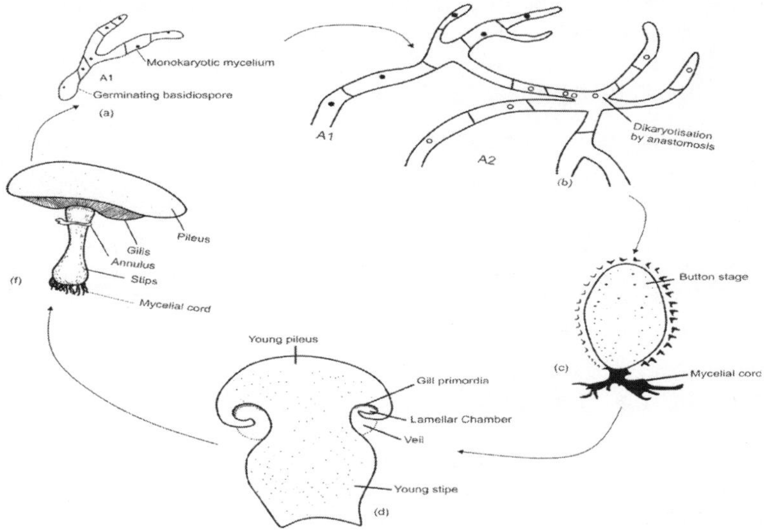

Figure 3.2: Life cycle of *Agaricus*

History of Button Mushroom Cultivation

Reports regarding the cultivation of mushroom is available, according to reports it was started in 1650s in France. Pure culture of mushroom was isolated in (1900), first time spawn was grown on grain in (1932) and good quality compost was standardized in (1950). Till the beginning of 21st century, the production of button mushroom was on number one. In India button mushroom is very popular in the market, in the name of mushroom people know only button mushroom. Button mushroom production can be done in controlled growing houses and under natural environmental conditions. Under natural conditions, it is a seasonal crop of winter. Seasonal growers use a variety of growing houses for cultivation of button mushroom *i.e.*, huts made up of bamboo, surkanda, grasses and straws, kuccha houses, poultry sheds and small godowns. Seasonal growers can be seen in the rural areas and outskirts of town and cities. Under controlled conditions round the year cultivation of button mushroom can be done *i.e.*, proper humidity, temperature, CO_2 concentration and fresh air can be maintained in the farm houses.

Nutritional Value of Button Mushroom

Nutritional value of button mushroom is very high. There are rich source of protein and recognized as a substitute of meat and it is cultivated worldwide. Mushrooms are rich in dietary fibers so good for digestion, water is 90.0 per cent, proteins 5.2 per cent, Fe 21.9 mg/kg, Cu 18.6 mg/kg, Zn 10.2 mg/kg, low in Sodium and high in Calcium contents. Phenolic compound are present in it *i.e.* Myricetin (2729.46 μg/gm), Caffeic acid 392.51 μg/gm, Gallic acid 280.45 μg/gm).

Fungi is known for their anti tumour activity as 1-3 β-D Glucan is present in it.

They contain carbohydrates, chitin, hemicellulose, β and α glucose mannans, xylans and galactans which may explain their beneficial effects on gut microbiota.

Production Technology

For the cultivation of mushroom basic knowledge of compost preparation, crop management practices in required. Cultivation of button mushroom include three Basic steps:

1. Composting
2. Spawning
3. Casing

1. Composting

Compost can be prepared by two methods 1) Long method 2) Short method. Seasonal growers generally use long method, which is old and single phase method. This method spreads upto 25-30 days, which include 2 days for pre-stack wetting of straw, followed by 8-10 turns.

Short method is two phase method. Phase-I. In this part compost is made outside in open area.

In Phase-II rest of the part of composting is done inside the special structure is known as tunnels.

Two main precautions are taken during compositing of straw. First the proper size of the compost stack *i.e.*, 150 cm x 150 cm is maintained for proper and thorough mixing of ingredients and moisture level upto 60-70 per cent. The dark brown colour of compost with a pH of 7.5 and without smell of ammonia are the basic features of good quality compost, which a mushroom grower seeks for the optimum productivity. Some industrialist use short method of composting to process, wheat straw and chicken manure is added in it. Compost is ready in 15-16 days, among that 3-4 days are taken for pre-stack wetting period (Figure 3.3).

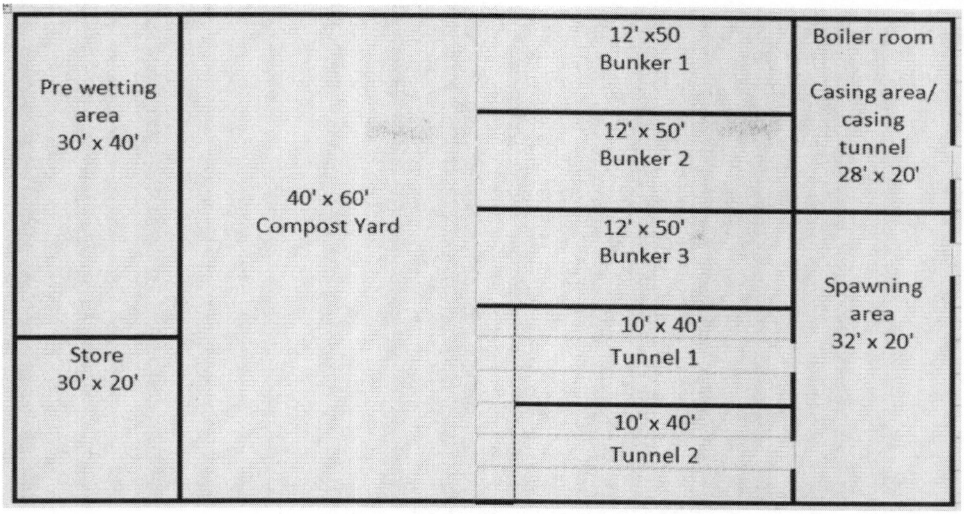

Figure 3.3: Structure of Composting Yard.

During out door composting, Phase-I is of 6 days and Phase II is of 6-7 days which is a peak time of heating and controlled fermentation of material. Initially temperature is maintained at 62-65°C for 8 hours, then it is lowered to 55°C and gradually drops to 45°C for 5 days, continuous flow of fresh air takes place, When elimination of smell of ammonia takes place that is the indicator of complete composting.

Ingredient

Compost can be prepared by adding straws, manure, gypsum, wheat bran, Molasses, Calcium, Ammonium nitrate. Carbohydrate material is the source of carbon than can be obtained from paddy, wheat, sugarcane bagasse, leaves of sugarcane, soyabean straw, cellulosic waste, corn cobs *etc.* Nitrogen can be obtained from brewer's grain, chicken manure and various bran/cakes or fertilizers. Nitrogen content can be balanced by adding Urea, ammonium fertilizers *etc.* During stacking proper C:N Ratio is needed, that can be adjusted to 25-30:, after the completion of composting which lowers to 16-20 per cent.

Compost Preparation by Long Method

This method is suitable for those conditions where area is not so large and small scale growing of mushrooms is there.

- ☆ The area where compost is to be prepared, thoroughly cleaning can be done by 2.0 per cent formaldehyde solution.
- ☆ Straw is spread uniformly on the sterilized floor, water is sprinkled over the straw by sprinkler or water pipe. Straw is turn frequently upside down to until it absorbs sufficient amount of water. Initially moisture in the straw can be upto 70-75 per cent.

- ✰ After 2 days when straw is completely wetted, it is arranged in a heap. Other ingredients like wheat bran, Chicken manure, gypsum, and other materials are mixed in it, water is sprinkled on it. All the ingredients are mixed properly and left for 24 hrs.
- ✰ All the ingredients are arranged into a high aerobic pile, upto height of 5 feet and 6 feet wide.
- ✰ Due to the activity of microorganisms, after 5-6 days as the reaction goes on in the pile, temperature rises to 70°C.
- ✰ Then mixing of piles is done after 6 days. It is turn upside, down and outside, inside turning is also there. So that whole of the mixture can be exposed to aerobic condition and homogenous. Wherever is required water is added into the pile.
- ✰ After 8-9 days as the composting of straw starts, light yellow colour of straw changes into the dark yellow/light brown in colour and there is a slight shrinkage in its volume. Some foul smell and production of ammonia also takes place from the heap.
- ✰ After 10 days, pile is opened and turned upside down and homogenous turning takes place. It further shrinks in amount. Temperature of heap becomes higher than before and colour changes into dark brown. More amount of ammonia is produced. Some powdery mass of Actinomycetes can be seen on the outer side of heap. It is the indicator of Good quality compost.
- ✰ After 12 days, pile is turned upside down and particular amount of Gypsum is added (3rd turning).
- ✰ After 15 days compost is turned and stack is formed (4th turning).
- ✰ After 18 days compost is turned and stack is made (5th turning)
- ✰ After 22 days compost is turned and stack is made (6th turning).
- ✰ After 25 days compost is turned and insecticide is added to make it insect free. (7th turning).
- ✰ After 28 days, pile is break open to check the smell of ammonia, if foul smell is not there, smell changes into sweet smell, is the indication that compost is ready to use for spawning. Spread the compost on sterile floor so that it becomes low in temperature.

Brain Tester

1. What can be the source of carbon in substrate formation?
2. What can be the source of Nitrogen during substrate formation?
3. Which material can be used for compost formation?
4. Role of Gypsum and Calcium carbonate during compost formation.

5. How C:N ratio is important for composting?
6. What are the qualities of good compost?

Innovative Idea

Initially farmers were preparing compost by the long method. There were chances of getting infection of fungi. Farmers start using bran or cakes and fertilizers as source of Nitrogen replacing chicken manure, later on they shift to short method instead of long method. In turning process also, they make changes in the row of compost. They start shifting compost into two parts so that maximum exposure to air takes place.

Zero Energy Poly Tunnel

In Solan, Mushroom Research Centre this technology was developed. During compost preparation for the button mushroom they insert pipes with holes in the compost pile. By this innovation, quality of compost was improved and farmers left the long method and switch on to short method for preparing compost.

Ingredients for Short Method

1.	Wheat straw	-	100 kg
	Wheat bran	-	5 kg
	Chicken manure	-	60 kg
	Urea	-	2 kg
	Gypsum	-	10 kg
2.	Wheat straw	-	100 kg
	Horse manure	-	5 kg
	Urea	-	1.5 kg
	Chicken Manure	-	90 kg
	Brewer's grain	-	10 kg
	Gypsum	-	15 kg

Ingredients for Long Method

1.	Wheat/rice straw	-	100 kg
	Urea	-	1.5 kg
	Cerium Ammonium Nitrate (CAN)	-	5.0 kg
	Super phosphate	-	1.0 kg
	Wheat bran	-	6.0 kg
	Molasses	-	1.5 kg
	Murat of Potash	-	1.0 kg
	Gypsum	-	10.0 kg

Features of Good Quality Compost

1. Compost is the basis for the growth of mushroom it should be of brown in colour.
2. It must be free from foul smell.
3. Compost should not be sticky or greasy.
4. pH of the compost should be 7.2 – 7.8.
5. Moisture content of the compost can be of about 65 to 67 per cent.
6. Compost should be free from molds.

Composting by Short Method

Now a days farmers and other growers are giving preference to short method. In this method all the ingredients are mixed properly, wetted and filled in the bunkers: through the pipes pressured air is introduced at regular interval so that aerobic conditions can be maintained.

1. Phase – I: This phase goes on the 10-12 days and outdoor composting is done.
2. Phase – II: It takes 7 days during this phase pasteurization and conditioning of compost takes place.

Phase – I

During I stage, composting is done in outdoor. Wheat straw and chicken manure is mixed and wetted properly. After 2 days again mixing of material *i.e.*, urea and wheat bran is to be added and turning of the material to be done upside down. Stack of the material is formed about 5^1 x 5^1. After 2-3 days again stack is breaked, gypsum is added in it and water sprinkled to maintain the moisture. After 7-8 days compost is ready for pasteurization. Compost seems to be brown in colour and moisture content to be maintained upto 75 per cent (Figure 3.4 - 3.8).

Figure 3.4: Wetting of husk

Figure 3.5: Spreading of husk

Figure 3.6: Bunker and tunnel for composting and pasteuraization

Figure 3.7: Mixing of Compost

Figure 3.8: Filling of compost in tunnel

Phase – II

This step takes place in tunnel, which is a insulated room with holes in floor to facilitate the circulation of air. Various steps takes place during this phase.

1. Pre-pasteurization conditioning.
2. Pasteurization
3. Post pasteurization

During first stage when compost is filled in tunnel, due to reaction rise of temperature takes place and the fresh air is supplied to maintain the level of air. Temperature is maintained till 45-50°C to avoid the growth of moulds. Ammonia also gets fixed in lignin – decomposition of humus take place and release of ammonia takes place.

During pasteurization the supply of fresh air is inhibited so that temperature can rise above 55°C. Purpose of this step to make compost microbes free. But temperature to be maintained upto 60°C above this temperature can be harmful for other necessary microbes.

During Post-pasteurization condition, temperature of compost is bring lower to 45-50°C. Enough oxygen is supplied to compost to maintain aerobic conditions. Pasteurization and supply of air make the compost suitable for the growth of button mushroom. At this stage compost is free from ammonia smell. Temperature is maintained at 45-48°C and fresh air flows through it and condition is known as Post pasteurization conditioning of compost (PoPC). After 3-4 days temperature is bring down to 25-30°C by introducing fresh air in tunnel. At this stage compost is ready for further use of spawning.

Compost is dark brown in colour, soft, free of foul smell, nitrogen content >2 per cent, Ammonia is below to ppm and moisture content is up to 65 per cent. (Figure 3.9-3.15).

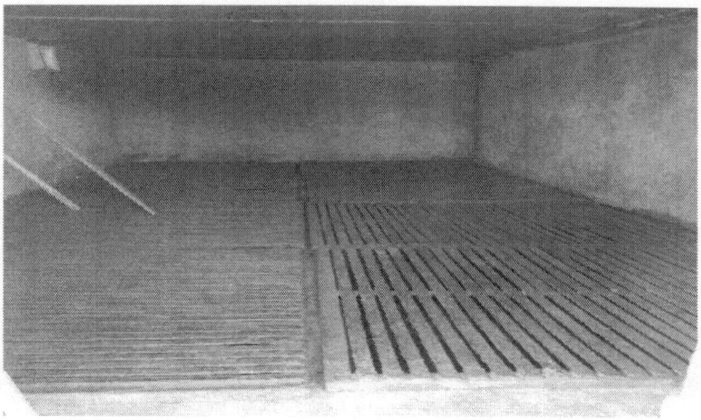

Figure 3.9: Tunnel with iron grating

Figure 3.10:Tunnel with pipes & spigots

Figure 3.11: Layering of husk in tunnel

Figure 3.12: Pasteuraization of compost at initial stage

Figure 3.13: Mixing of Compost with help of machines

Figure 3.14: Composting process in tunnel

Figure 3.15: Piling of compost at final stage

Spawning

Grains with vegetative mycelium of fungi is known as spawn. Grains can be of wheat, sorghum and bajra. Spawn can be procured from Laboratories, seed suppliers for small scale industries large scale industries can manage their own lab for the spawn. Seasonal growers can procure it from there nearby laboratories. Many agricultural universities are working on it.

1. Directorate of Mushroom Research Centre (Solan)
2. G.B. Pant University of Agriculture
3. IARI Delhi
4. C.C.S. Haryana Agricultural University, Hissar
5. Punjab Agricultural University, Ludhiana.
6. National Research Center for Mushroom, Solan.
7. Punjab State Department of Horticulture at Jalandhar and Patiala.
8. Maharana Pratap Horticulture University, Murthal, Sonepat
9. J&K State Department of Horticulture at Jammu and Kashmir.

Variety of strains of *Agaricus bisporus* are in use for the production purpose, which claimed to be highly productive and suitable for local environment.

- ✰ Strain S11 of *Agaricus bisporus* was frequently used all over the region.
- ✰ Two hybrid strain U_1 and U_{-3} from Holland are common for the growing purpose.
- ✰ Farmers of Punjab, Haryana and J&K are using the strain MS39,l which is introduced from Ludhiana.

- Strain P_1 from National Research Centre for Mushroom (NRCM) Solan is suited for long method compost.
- Three productive strain's NCS 100, NCS 101 and NCH 102 of *A. bisporus* have been released by NRCM, Solan for cultivation in the country.

After çooling of the compost, spawn can be added in it at the rate of 0.5-0.75 per cent and compost with spawn can be used/filled in three systems *i.e.*, wooden boxes, polybags and tiered beds. Farmers and growers in hilly regions generally used polybags for growing mushrooms. While in plains farmers and growers spread the compost in beds up to depth of 4 to 6 inch and do the spawning on beds itself.

In bags generally 10-12 kg compost is filled, farmers also uses the wooden boxes or empty fruit crates for growing mushroom or they lay beds on the shelves created temporarily from bamboo and polysheets. For permanent structure it can be fabricated of steel or iron structure. A gap of minimum 50cm is left between the bags. Bags are kept at a temperature of 23-24°C and temperature inside the bag is higher ± 2°C. As the mycelium starts spreading inside the compost, there is a rise in the temperature. During running of the spawn in the compost, rooms are generally kept closed,CO_2 level can go up to 10,000 ppm, which can be suffocating process. Running of spawn on whole surface takes 12-15 days, it turns white in colour. At this stage our bags/beds are ready for casing. (Figure 3.16-3.21).

Figure 3.16: Filling of Bags

Figure 3.17: Bagging

Figure 3.18: Racks for keeping the Bags

Figure 3.19: Incubation of Bags

Figure 3.20: Incubation

Figure 3.21: Mycelium Running on Compost

3. Casing

It is a process by which the covering of compost is done with a thin layer of soil or soil like material, after the spreading of spawn on the compost. Casing is done due to many reasons.

1. It provides humidity as casing soil can hold water for a longer period.
2. Casing soil provides strength to the fruiting body by providing nutrient.
3. As soil can hold moisture it prevents quick drying of the compost, it will help in better growth of spawn.
4. Casing soil helps in the maintenance of temperature, if temperature will fall then fruiting of mushroom will be disrupted.

5. Fruiting of mushroom, from vegetative mycelium takes place on nutrient rich medium.

For casing, about 1.5-2.0 inch thick layer of casing soil is to be spread over the compost layer, which is with fully run mycelium. Casing soil can be the mixture of burnt rice husk, farm yard manure, coir pith and sandy loam soil. Casing soil should not be in the powder form or sticky otherwise lumps will be formed and casing layer will be jam-packed and fruiting will not takes place. Water will drench and nutrient will deplete from it. For casing, garden loam soil and sand soil can be added in 4:1 ratio or it can be loam soil and decomposed cow dung in 1:1 ratio, sand and lime are commonly used and 2-3 years old spent compost can also be used.

Figure 3.22: Casings soil on the Bags

Figure 3.23: Casing soil on basket

Cropping

After spreading the casing soil on the bags/beds, moisture is maintained by watering. For a week temperature is maintained around 25°C during this period, mycelium travels upward towards the casing soil. Then fresh air is introduced to lower the level of bags, carbon dioxide and temperature is maintained at 1500 ppm and 16-18°C respectively. These conditions helps in the induction of small pinheads on the upper side of casing layer, with in 3-4 days pin heads mature in a full fruiting body (Figure 3.24-3.27).

Figure 3.24: Bed of Button Mushroom

Figure 3.25: Bags of Button Mushroom

Figure 3.26: Fruiting of Button Mushroom

Figure 3.27: Harvesting of Mushroom

- Temp – 16-18°C
- CO_2 Conc – 1000-1500 ppm
- Humidity – 85-95 per cent

In general 3-5 flushes can be obtained within 4-6 week time. When mycelium runs in the compost more CO_2 is produced. The level of CO_2 to be lowered during the fruit formation, otherwise stipe part – will be longer in size.

Composting

↓

Spawning at the rate of 0.5 per cent

↓

Bags are shifted to incubation room for run of spawning

Temp 25°C | No Fresh air

Humidity 85 per cent | 11-14 days

↓

Add casing soil on the top of bag/bed after complete spawning.

2-10 days | No fresh air

Humidity – 85 per cent

Temp-25°C

↓

On the casing soil white mat of fungal mycelium can be seen

Temp – 16-18°C | 7-10 days

Humidity – 90°C

CO_2- 1000-1500 ppm

↓

Small Pinheads appears among the white mycelium

Temp – 18°C | 5-6 days

Humidity – 85 per cent

CO_2- < 1500 ppm

↓

First Flush comes after 3-4 days

Temp – 16-18°C | 7-10 days

Humidity – 85-90 per cent

CO_2- < 1500 ppm

↓

Second flush again after 5-6 days

Packaging and Processing

As the crop of mushroom is ready to harvest growers and farmers prefer to harvest the fruiting bodies in night, in the morning they can go to wholesale market for the supply. Time gap between picking and supply to be shortened to avoid the spoilage. Shelf life of mushroom is small. If time gap between picking and sale will be less than browning of fruiting bodies can be avoided and good price can be fetched from it, profit will be more. After harvesting the crop can be kept at low temperature and in refrigerated van crop can be transported to the wholesale market. Packaging of mushroom can be done in small boxes. Generally picking of mushroom is done only once a day, but if demand is high in the market then it can be done twice in a day. Unwashed mushroom are generally preferred as they retain the quality of mushroom but soil particles attached to the sporophore gives the unhygienic look. So preference is given clean and white mushroom which gives a aesthetic feel. The processed mushrooms can be stored for further use for making pickels and other products. There can be long term storage and short term storage of mushrooms (Figure 3.28).

Figure 3.28:After harvesting of Mushroom

Economics

As button mushroom is a winter crop, it can be grown in natural conditions and under controlled conditions. If it is grown in natural condition then our investment will be less but under controlled conditions investment will be more and net profit will be less.

Analysing

Cost of production depends upon the biological efficiency and number of flushes. While analysis the cost of production per kg in commercial units, there is a wide variation due to differences in number of flushes taken per year and biological efficiency. In one commercial unit cost/kg is Rs. 65 at 18% BE & G flusher per year another grower have taken 4 flushes and BE is much below the economics level. So, during Calculation the cost of production per kg mushroom it is must to mention the BE and number of flushes harvested. In Developed countries BE is about 32-35 per cent, due to better use of technology. The cost is mushroom is very less and yield is about double. To reduce the cost of product, BE is to be increased and number of flushes to be more at the same time. Cost of mushrooms depends upon the BE and the number of flushes produced. Seasonal growers take only one crop, BE varies from 15-25 per cent. The larger BE in due to the fact that growers while taking only one crop, they keep on appearing. Due to disease BE goes down and that affects the flushes of crop and cost of the crop.

Cost Benefit Analysis

Basically cost of production depends the type of hut, machinery, spawn, Raw material and labour. Cost of seasonal growers is less as they are using natural condition, do not using electricity. They are not spending on land, in one corner of field they can grow it, rental value is not added. Most of the commercial grower get BE about 22 per cent. Various components plays a role in it (Table 3.1).

Compost, Labour and electricity account for 2/3rd of the production in commercial cultivation. It is important that the compost of good quality is produced to ensure proper yield and low cost of production. We need better technology and environment control to produce more and lower the cost of production. There is also need for rationalisation of electricity changes vis-a-vis other near by competing states. Farmers producing by short method have to invest in tunnels for pasteurization, get compensable by less cost on labour, more compost per kg of raw material higher and stable yield. Its one time investment but one can get benefit for many years.

Table 3.1: Contribution of different Components on per kg Production in Short Method and Long Method at BE 18 per cent and 22 per cent

Particular	*Commercial Growers Short Method Composting*	*Seasonal Growers Long Method Composition*
Biological Efficiency	18 per cent	22 per cent
Room and machines	10.70 per cent	5.50
Spawn	4.30	2.75
Raw material	24.60	33.50
Labour	13.75	15.25
Electricity	11.65	-
Total	65.00	57.00

Source:- Report Haryana Kisan and Agricultural Cost and Prices Commission (2019).

Short method of composting has better economic and environmental benefits and is more sustainable due to high assured yield with less disease incidence by short method. Compost hubs are available which provide ready-made compost. The cost of production of oyster and milky mushrooms is very less than button mushrooms.

Market System

There is no organized market for mushroom. Depending upon the amount of production, farmers sell it in local market, metros or big cities.

- ☆ Some farmers enter in to a contract with canners and sell mushrooms at fixed prize.
- ☆ Most of the farmers sell their fresh produce in Azadpur (Delhi).
- ☆ Depending upon the need and prevailing prices the mushrooms are also sent for sale to cities like Patiala, Kanpur, Meerut, Jaipur, Chandigarh, Jammu, Shimla, Mumbai, Kolkatta, Raipur, Dehradun, Lucknow *etc.*
- ☆ There are collection centers like Reliance fresh, Easy day, Mother dairy *etc.*
- ☆ Very small amount of mushroom is used for pickling by agencies like (Pachranga, krishna Pickels, Home pickle, Haic pickles and Rajesh Pickels *etc.*)

Pathogens Associated with Button Mushrooms

Many fungal, bacterial, insect – pest and nematode infection can be seen in the compost. The common fungal include brown plaster and false truffles which are common at higher temperature during October or after February.

1. Brown Plaster (*Populospore byssina*)
2. Inky caps (*Coprinus* sp.)
3. Bacterial Blotch (*Pseudomonas tolassii*)
4. Cobweb (*Dactylium dendroides*)
5. Saprophytic (*Aphelenchoides* sp.)

Fungicidal spray helps in checking the problem of fungi and bacteria, spray of chlorinated water also helps in preventing the disease.

Growing of button mushrooms is an profitable venture. Many people are in this business from many years and earning profit of good sizeable amount. Quality mushroom in the market can fetch lucrative prices in various regions of countries. For beginners it can be a small scale industry. But those who are in this business can be a large scale industry can earn profit by supplying in market or by drying. Canning and processing can be exported. That will help in improving the economy of country.

QUESTION

1. **Discuss about classification of Button Mushroom.**
2. **Explain through flow chart long methods of composting.**
3. **What is casing, mention its significance?**
4. **Nutritional value of Button mushroom.**
5. **Mention some fortified spices of Button Mushroom.**
6. **Discuss various method of preservation of Button Mushrooms.**
7. **What precautions to be taken while drying the Button Mushroom?**
8. **Define canning freezing and other method of preservation.**
9. **How entrepreneurship can be established in this field?**
10. **Discuss about the problems faces by farmers during composting.**

Chapter 4

Cultivation of Oyster Mushroom (*Pleurotus* sp.)

Classification

Kingdom	–	Eumycota
Phylum	–	Basidiomycota
Class	–	Agaricomycetes
Order	–	Agaricales
Family	–	Pleurotaceae
Genus	–	*Pleurotus*

Introduction

Oyster mushrooms are cosmopolitan, and belong to the genus *Pleurotus*, class Basidiomycetes and family Pleurotaceae, popularly it is known as Dhingri in India. It grows naturally in the temperate and tropical forests on dead and decaying wooden logs or sometimes on dying trunks of deciduous or coniferous woods. It can also grow on decaying organic matter.

BRAIN TEASER

Q.1. What is the popular name of *Pleurotus* ?

Q.2. Why Pleurotus is known as *Pleurotus* ?

Q.3. When the cultivation of *Pleurotus* starts and in which country ?

Origin

First time it was cultivated in Germany by Flack as a subsistence measure during World War-I in 1917. Now it is grown commercially around the world for food. Oyster mushroom can be used industrially for mycoremediation purposes, for making green leather and biofuel.

The fruit bodies of this mushroom are distinctly shell or spatula shaped with different colours of white, pink, yellow, blue, cream, grey or light brown depends upon the species.

Nomenclature

Both Latin and common names refer to the shape of the fruiting body. The Latin *Pleurotus* refers to the side ways growth of the stem with respect to the cap, while the Latin *Ostreatus* refer to the shape of the cap which resembles the bivalve of the same name. Many also believe that the name is fitting due to a flavour and resemblances of oyster.

Description

The oyster mushroom have three distinct parts – a fleshy shell or spatula shaped cup (Pileus) (about 5–25 cm in diameter 1.9–7.8 inches in length). Fleshy, with eccentric or lateral stipe; colour may be white, cream, yellow, pink, gray or browish. A short or long lateral or central stalk called stipe and long ridges and furrows underneath the Pileus called gills or lamellae. The gills stretch from the edge of the cap down to the stalk and bear spores. The gills of the mushroom are white to cream, and descend on the stalk, if present. The stipe is off-center with a later attachment to wood. The spores are smooth, cylindrical and germinate very easily on any kind of mycological media with in 48 hrs. The spores print of the mushroom is white to lilac-gray, and best viewed on dark background. The mycelium of *Pleurotus* is pure white in colour (Figure 4.1).

Q.1. Why Oyster mushrooms known as primary decomposers?

Q.2. On which substrates it can be easily grown?

Q.3. Mention the structure of fruiting body.

Q.4. Is there any colour variations in the fruiting bodies of oysters.

Habitat

Oyster mushroom is widespread in many temperate and subtropical forests through out the world. It is a saprotroph that acts as a primary decomposer of wood, especially deciduous trees, and beech trees in particular. It is a white rot wood-decay fungus (Roger, 2006). The oyster mushroom is one of the known carnivorous mushrooms. Its mycelium can kill and digest nematodes, that is way in which the mushroom obtains nitrogen. The standard oyster mushroom can grow in many places, branched oyster mushroom, grow only on trees. They may be found all year around in the UK. Oyster mushrooms benefit the forest by decomposing the dead wood, returning vital elements and minerals to the ecosystem in a form usable to other plants and organisms and can be the part of sustainability (Stamets, 2000).

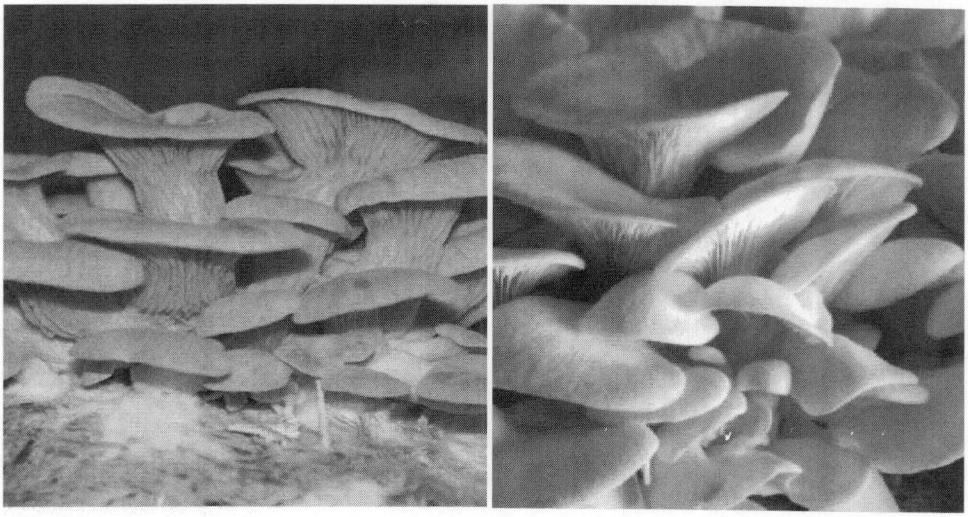

Figure 4.1: Fruiting bodies of *Pleurotus.*

First large scale cultivation on logs was achieved in Hungary in 1969. Later a variety of lignocellulosic by products from agriculture or foresting were also reported to be good substrates and various species were brought into cultivation throughout the world.

Such as grey oyster mushroom (*P. sajor-caju*) tree oyster (*P. ostreatus*), white oyster mushroom (*P. florida nomen nudum*), abalone mushroom (*P. cystidiosus*), golden oyster mushroom (*P. citrino pileatus*), pink oyster mushroom (*P. flabellatus*) black oyster mushroom (*P. sapidus*) *etc.* At present it is the second most important cultivated mushrooms in terms of world production (Carrera, 1998) (Figure 4.2a-g).

Taxonomy of *Pleurotus*

As *Pleurotus* are adapted to a broad range of environmental conditions and substrates, and have a extraordinary genetic diversity has placed taxonomists in a demanding situation. Fruit body morphology, microscopic observations, mating studies between populations, biochemical analyses have not provided clear-cut results. Through molecular studies, intra and inter specific heterogeneity was determined using ribosomal DNA sequences indicated geographic speciation in several groups. But, many indigenous populations have some commercial strains and widely distributed through the world and escaped from cultivation. Taxonomy and systematics *Pleurotus* sp. is still far from being solved. It requires not only the identification of effective classical and molecular characters, but basic consensus on speciation processes and species concepts with in the genus.

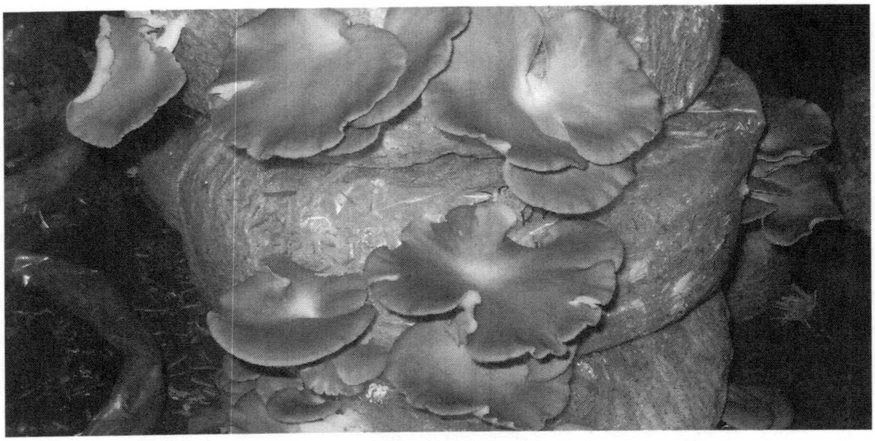

Figure 4.2a: *P. sajor-caju*

Figure 4.2b: *P. florida.*

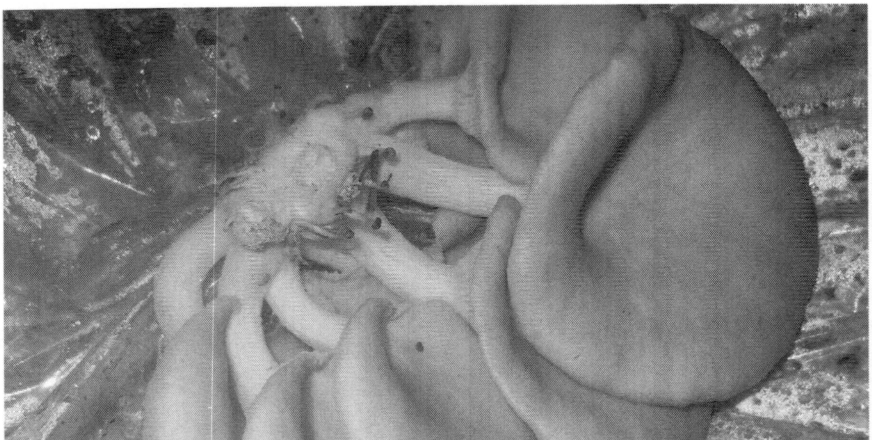

Figure 4.2c: *P. columbinus*

Figure 4.2d: ***P. djamor***

Figure 4.2e: ***P. ostreatus***

Figure 4.2f: ***P. flabellatus***

Figure 4.2g: ***P. citrino-pileatus***

Life Cycle of *Pleurotus* (Oyster Mushroom)

The life cycle of oyster mushroom is typical as of Basidiomycets. It begins with the germination of a basidiospore in a suitable substrate. On germination of basidiospore monokaryotic mycelium formation takes place which contains genetically identical nuclei(n) and capable of indefinite independent growth. When two compatible monokaryotic mycelia comes into contact then fertile dikaryon hyhal fusion takes place by plasmogamy (n+n) clamp connection formation takes place and binucleate hyphae contains two genetically different nuclei (+ and -) strain through out the mycelium. When the environmental conditions such as humidity, temperature, light are appropriate, fruiting body formation takes place. From the pin head to fully matured body with in 4-5 days formation takes place. Basidiospore formation takes place on the basidia, which are club shaped binucleate cells, formed in the lamellae (hymenium) of each fruit body, karyogamy and meiosis takes lace. The four haploid nuclei move to the sterigmata on the basidium, to form four new basidiospores. When the fruit bodies are mature basidiospores are discharged, starting the sexual life cycle, again. In species *P. cystidiosus* asexual cycle is reported in which small structure formation takes place known as synnemata (1-10 mm height; 0.03-0.39 inch in diameter) (Figure 4.3).

Know Yourself

Q.1. Spores in *Pleurotus* is homokaryotic or heterokaryotic in nature.

Q.2. Define Synnemata and when its formation takes place.

Q.3. Discuss various step in the cultivation of *Pleurotus*.

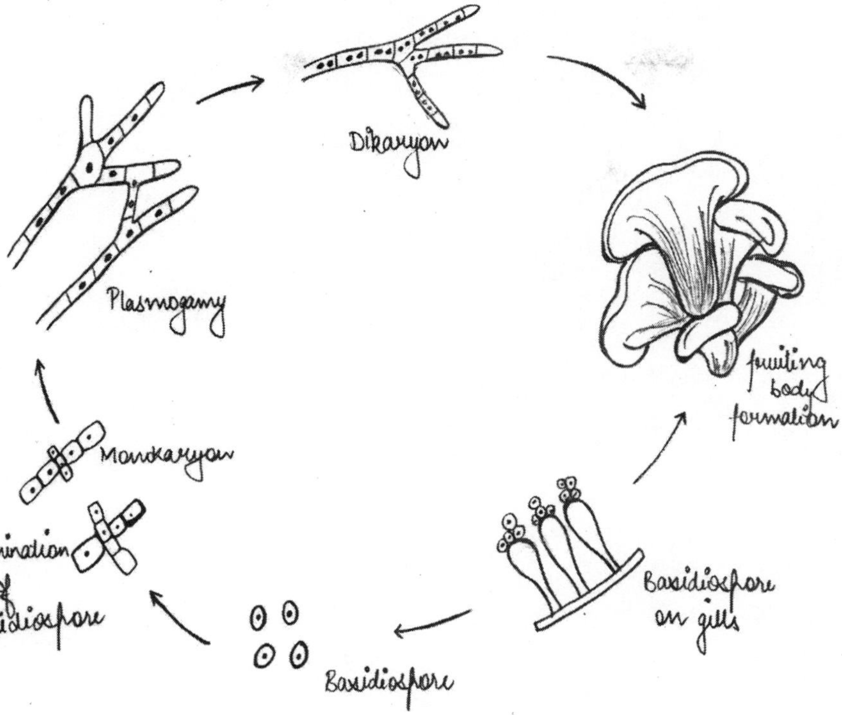

Figure 4.3: Life cycle of *Pleurotus*

Cultivation Technology

The cultivation of oyster mushroom can be divided into four steps:

1. Preparation of spawn
2. Substrate preparation
3. Spawning of substrate
4. Crop management

Spawn Preparation

A pure culture of *Pleurotus* sp. is needed for inoculation on sterilized substrate,. It takes 10-15 days for mycelial growth on cereal grains. With a range of fruiting temperatures (15°-30°C, 59°-86°F), commercially various strains of oysters are available. These days for spawn preparation main cereal grains used are (wheat, millet and sorghum). These grains are sterilized in glass jars or PP bags then inoculated with a selected strain. The inoculated cereal jar or PP bags are incubated at appropriate temperature for complete colonization. In an alternative method, basidiospore prints serial dilution can be made and grain spawn can be prepared from it (Figure 4.4).

Figure 4.4: Grains

Substrate Preparation

Before substrate preparation several things can be done: Selection of raw materials, storage of raw materials, cutting of raw materials.

a) Selection of Raw Materials

For the production of oyster mushroom straw is the basic organic material. It can be obtained from a variety of crops such as rice, wheat, rye, barley, oat, peanut hull, buckwheat, sunflower *etc.* Straw should be dry, free of mold and moisture content should not exceed to 12 per cent. To boost production of mushroom, gypsum, limestone, sugar, green hay, alfalfa, wheat, rice can be added and mixed. Sawdust can also be used in place of straw (Figure 4.5).

Figure 4.5: Straw for Composting

b) Storage of Raw Material

Raw material should be stored properly kept dry to avoid mold formation and loss of materials. Raw material can be placed outside covering with the trampoline to maintain the quality organic material or it can be stored in sac bags (Figure 4.6).

Figure 4.6: Straw is filled in bags.

c) Cutting of Raw Materials

Next step is chopping of the straw. The collected straw are cut into small pieces usually the length of 5-10 cm and soaked in fresh water for 8-16 hours. Straw are light and less distributed organic material even after soaking and boiling in hot water there is not so much change in their density. For the good growth of mushroom, density of straw must be proper, if straw is small in size more moisture will be retained. If the size is too big there will be pore and the mycelium cannot be there that will help to develop nematodes and other insects in it. Cotton seeds hulls and sunflower husk can be used to grow commercial production of mushroom.

Preparation of Substrate

Oyster mushroom can be cultivated on a large number of agrowaste having cellulose and lignin. It helps in more enzyme production of cellulose that is helpful in more yield. These include paddy, ragi, and wheat stalk, leaves of maize, millets and cotton, sugarcane bagasse, saw dust of hardwood, jute and cotton waste, dehulled corncobs, dried grasses, sunflower stalks, used leaf waste and peanut shells. It can also be cultivated by using industrial wastes like paper mill sludges, coffee by products, tobacco waste, *etc* (Figure 4.7).

Figure 4.7: Spread washed husk on the table

Substrate can be sterilized by the following methods:

1. Hot water treatment
2. Steam pasteurization
3. Sterile technique
4. Fermentation or composting
5. Chemical sterilization

By soaking the straw in hot water, helps in killing the pest and microorganisms present on the straw. Pasteurization also helps in making the straw free from microbes.

Sterilization

It is a process of killing microorganisms which are harmful to mycelium growth. Heating temperature depends on the quality of substrate. If the quality is good, sterilization temperature is maintained in between 95-99 degree Celsius and the substrate is heated for about 4-5 hrs. If sand or dirt is present in substrate, then heating can be done at 60°C (Figure 4.8a).

Pasteurization

For small scale production, oil drums are used for pasteurization. Steam is used in this process for making the straw microbes free. The drums are coated or formed a good compartment inside to avoid burning of bags. The drums is layered into two half vertically or 2-3 half horizontally. Lower half of the drum is filled with water and upper half of the drum is filled with bags. After placing the bags inside the drum, lid is closed tight from the top to build sufficient pressure. The heating time depends upon the size of the bags, usually it is heated for about 4-5 hrs. Then the bags are cooled down to 25°C (Figure 4.8b).

Substrate can be bagged before and after pasteurization depends upon the method used for sterilization. Addition of Gypsum powder 1 per cent, lime powder 2 per cent, compound fertilizer 2 per cent reduces the microbes or kills the microbes present on the straw. After preparation of substrate, it is compressed into the PP bags or wooden frame. Standard size of PP bag is 20 x 30 cm, 18 x 25 cm, wooden frame can be of 45 x 30 x 15 cm.

Figure 4.8a: Sterilization of Compost straw.

Figure 4.8b: Sterilization of Straw by boiling in hot water.

Spawning

It is a process in which mycelium of fungi is transferred to bags filled with substrate. It can be purchased from farmer shops or from spawn suppliers. In villages one can found free of cost from seed shops as Govt. of India is promoting the mushroom production (Figure 4.9).

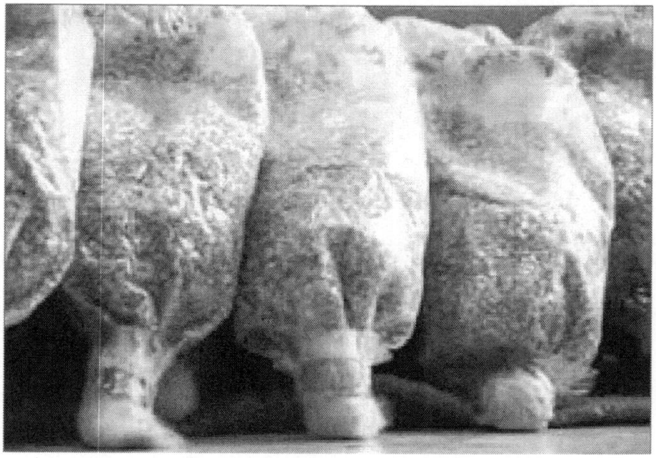

Figure 4.9: Commercially available Spawn

Inoculation

To avoid contamination of microbes this step to be done in a clean room. The person who perform this task must sterilize his/her hands or wear gloves. About a tea spoonful of spawn is placed in the substrate. Inoculation can be done in two ways. First one is thorough mixing of inoculum to the substrate about 10 per cent (w/w). Second one is layering process in which the substrate or straw is filled in the polypropylene bags and one tea spoon full of spawn can be placed in the periphery of the bag. 3-4 layer can be made in the bags. After inoculation bags should be packed tightly and small holes can be made by sterile knife or scissors (Figure 4.10a).

Figure 4.10a: Filling in the Bags.

Incubation

After inoculation bags with mycelium of oyster mushroom, bags are placed in dark incubation room, where the temperature are optimized to 20-25°C for the best results. They are kept in incubation room for about 15-20/25 days at 20-25°C temperature (Figure 4.10b-e).

Figure 4.10b: Placing of Bags in racks.

Figure 4.10c: Bags can be placed on floor.

Figure 4.10d: Fully grown mycelium in Bags.

Figure 4.10e: Fully grown mycelium in Bag

The incubation of bags with substrate depends upon the size, material and condition of the bags. These bags can be placed in various ways. Horizontal trays, shelves, vertical plastic sacks and pressed rectangular blocks can be used. Touching of bags from each other should be avoided. As soon as the white mycelium spreads on the substrate, PP bag with substrate can be moved to the production room. In some cases same room can also be used for production. During fruiting of body 10-12 hrs. of light is required in the room. During fruiting stage proper humidity, CO_2 and water is maintained. Mushroom contains about 90 per cent of water thus maintaining humidity of the atmosphere is important during the early stage of fruiting. Frequent watering about 5-6 times during the reproductive growing phase is required. Supplying of water directly to the mushroom should be avoided. Parameters to be followed for the oyster mushroom inoculation and fruiting shown in the Table 4.1.

Table 4.1: Various Parameters Required for the Growth of *Pleurotus*

Sl.No.	*Parameters*	*Spawn Phase*	*Pin Head*	*Fruiting Body*
1.	Temperature	20-25°C	10-15°C	12-20°C
2.	CO_2 (ppm)	5000-20,000	500-1000	600
3.	RH per cent	90-95 per cent	95-100	85-90
4.	Light (lux)	-	500-1000	500-1000
5.	Air replacement	1	4-8	4-5

Harvesting

Production of fruit bodies varies according to each species, spawn quality, substrate quality, environmental conditions and impact of pests (flies, mites) and diseases. Average biological efficiencies (yield of fresh mushroom as a percentage of the dry weight of substrate at spawning) reported from diverse substrates range from 35-59 per cent, considering a whole production cycle of about 70-80 days. The mushroom can be ready to harvest in approximately 3 to 5 days after the first mushrooms begin to form. When the cap of the largest mushroom in the cluster begins to go from convex to concave, pick the whole cluster. Blue oyster and elm oyster mushrooms grows in clusters of multiple mushrooms. It is generally best to pick the entire cluster when the largest mushroom in the cluster ready to harvest. A white powder on the side of the bag means the mushrooms are simply dropping their spores. Margins of the mushroom curls upwards is also the indication of maturity and ready to harvest (Figures 4.11 and 4.12).

Answer the Following

Q.1. What type of environmental conditions are required for the mycelium to run on the substrate ?

Q.2. How one can come to know that fruiting body is quite mature to harvest?

Figure 4.11(a & b): Initial pinning of fruiting bodies.

Figure 4.12: Various Stages of fruiting bodies.

Post Harvest Management

Storage

Short Term Storage

Freshly harvested oyster mushrooms can be stored at low temperature of (0-5°C) for 1-2 weeks without the loss in quality. Mushrooms can be packed in paper bags with inner lining, if it is to be sent to the distant markets (Figure 4.13).

Figure 4.13: Packaging

Long Term Storage

Dried mushroom with 2-4 per cent moisture, can be stored for 3-4 months in sealed pouches without any change in colour and taste. The dried produce can be rehydrated in lukewarm water (40-50°C) with in 20-30 minutes giving 80-90 per cent of original weight (Figure 4.14).

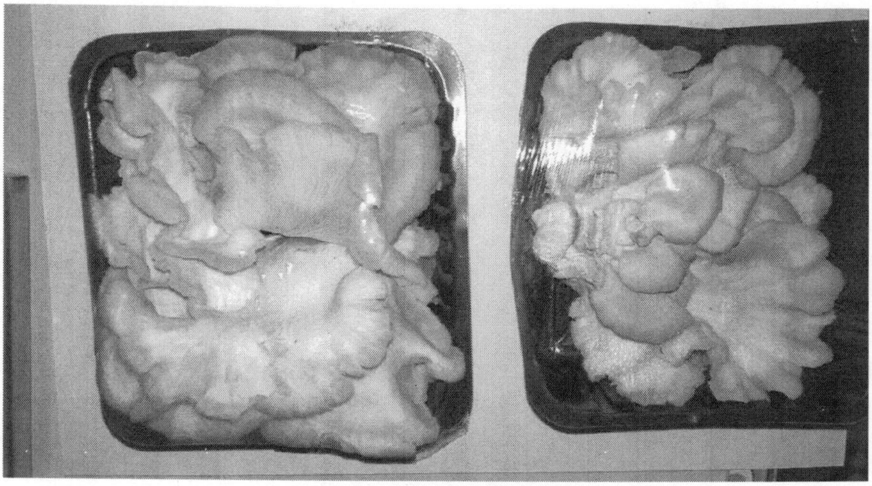

Figure 4.14: Packaging

Packaging and Transportation

Fresh mushrooms are packed in perforated paper bags/polythene bags. Poly pouches containing crushed ice and over wrapped in paper are put in trays/baskets which are then covered with thin polythene sheet with sufficient perforation for proper aeration. The prepacked pouches (250 or 500 gm) can be transported by roadways in trucks, buses depending upon the quantity to be transported.

Marketing

Domestic marketing does not face a problem at present because only small quantity are being streaded. As production develops, marketing promotion measures will need to be undertaken to meet the demand.

Export potential exists and needs to be taken advantage of by organizing cooperatives of producers linked to commercial units for processing fresh mushroom into dehydrated powder for export. A processing technology is fundamental for commercial production of oyster mushrooms undergo rapid-post harvest deterioration at room temperature. Canning, drying can be carried out depending on marketing strategies. After mushroom cultivation, spent substrates can be recycled as organic fertilizer, animal feed, substrate for other cultivated mushrooms, as well as for biogas, or enzyme production, paper manufacture and production of cardboard.

Do You Know This

Q.1. What can be the method of storage of oyster mushroom ?

Q.2. Is there any short term or long term method for storage ?

Q.3. What precautions to be taken while transporting fresh oyster mushrooms?

Q.4. Mention the Industrial use of *Pleurotus.*

Hygiene and Market Statistics

Oyster mushroom is known as the power house of nutrition. It contains moderate amount of fibers, minerals, Vitamin B and D, copper, iron, potassium, magnesium and low in fat and calories, 100 per cent cholesterol free. It contains antioxidants which is believed to prevent cancer.

The mushroom made up $ 1.22 billion of revenue for the U.S. economy in 2016-17 and increased up to $ 50,034.12 million in 2019 (Mahesh and Rohan 2018-2019).

The global oyster mushroom cultivation market size grow from $2.69 billion in 2022 to $2.85 billion in 2023 at a compound annual growth rate of 6.1 per cent. The Russia Ukraine war disrupted the chances of global economic recovery from the covid 19 pandemic, at least in the short term. The war between these two countries has led to economic sanctions on multiple countries a surge in commodity prices and supply chain disruption, causing inflation goods and services and affecting many markets across the globe. By 2027 the market size of oyster mushroom cultivation is expected 10 $ 3.55 billion at a CAGR of 5.6 per cent.

World Production

The cultivation of oyster mushroom has spread out rapidly world wide, as a natural result of their remarkable biological features and the versatile technology available. In general two main trends can be identified.

1. Private commercial enterprises operating primarily on a large or small scale for accumulation of capital and involving intensive cultivation tending to be highly mechanized.
2. Rural production for satisfying regional needs, which is performed through simple cultivation methods.

Developing countries have benefitted by the latter trend in terms of its local contribution to food production, rural development and sustainable agriculture. United States, China, South Korea, Japan and Indonesia are major producer of it and at present *Pleurotus* has become the second most important cultivated mushroom, accounting about 22 per cent of total world production.

Culinary Uses

The oyster mushroom is frequently used in Japanese, Korean, Chinese cookery as a delicacy. It is served in the form of soups, stir fry recipes with soy sauce or in stuffed form. Bread, pizzas and other bakery products can be prepared by the use of *Pleurotus*. It has bitter sweet aroma of benzaldehyde which is characteristic of bitter almonds. It has a odour like anise and in taste is mild. Oyster mushroom and young stage is best as it becomes aged it becomes tough and flavour also changes.

These are widely cultivated and used in Kerala, India where a wide variety of dishes are prepared from it. Oyster mushrooms are main cultivated in large PP bags with many layered in the bags and spawn is placed between the layers.

Oyster mushroom contain small amounts of arabitol, a sugar alcohol, which may cause gastrointestinal upset in some people.

Economic Importance

Economic importance of mushroom lies in its uses as food for human consumption.

It is rich source of Vitamin C and B complex and protein content varies between 1.6 to 2.5 per cent.

It has most of the mineral salts required by the human body. The Niacin content is about ten times higher than any other vegetables.

The folic acid present in oyster mushroom helps to cure anemia.

It is useful for the patients of hyper-tension, obesity and diabetes due to its low sodium; potassium ratio, starch, fat and calorific value.

Alkaline ash and high fibre content makes them suitable for consumption, persons suffering from hyperacidity and constipation.

A polycylic aromatic compound pleurotin has been isolated from *P. griseus* which possess antibiotic properties.

Spent straw can be re-cycled for growing oyster mushroom after supplementing with wheat, rice bran @ 10-15 per cent and for preparing compost of white button mushroom after suitable supplementation with Nitrogen rich source, chicken manure.

Spent straw can be used as cattle feed or by bio-gas production. Slurry can be used as manure.

Answer the Following

1. **Mention the economic importance of *Pleurotus*.**
2. **What are the bioactive compounds present in it?**
3. **Can oyster mushrooms be used to cure any disease?**
4. **What can be the harmful effect of *Pleurotus* ?**

Disadvantages

Oyster mushroom release enormous amounts of spores into the atmosphere of growing rooms. The inhalation of these spores induces allergic reactions in about 30-40 per cent of farm workers. Spore allergy is characterized by influenza like systems which disappear without treatment in a few days if exposure is prevented. It can be controlled by using masks and having efficient ventilation system (M. Carrera, 1998).

Disease and Management on Crop

- The crop is attacked by flies, sprint tails and mites. Timely spraying with insect specific insecticides is recommended.
- Several fungal competitors such as *Aspergillus, Cladosporium, Fusarium and Rhizopus* sp. have been found on the substrate. Spraying with Benomyl and Bavistin is recommended to control the disease.
- Crop is subject to diseases like yellow blotch, brown spot and bacterial rot, control measures are needed to include.
- Proper management of temperature and humidity during growth period.
- Regular application of chlorinated water containing 100-150 ppm of freely available chlorine at an interval of 3-5 days.
- Application of oxytetracycline and streptocycline.

Economic Analysis of *Pleurotus* Growth

Cost of Plan

Mushroom	Oyster
Substrate	Straw
Hut (room)	1 (25' x 60' x 14')
Bags in each room	(1000 (2 kg in each bag dry)
Biological efficiency	85 per cent
Total Produce (Crop)	1700 kg

a) Room in which crop to be grown

Sl.No.	Material	Quantity	Measurement	Rate	Amount
	Room		(25' x 60' x 14')		
1.	Bamboo	50	30 x 3"	300	15000.00
		15	25 x 1"	600	9000.00
		6	1.5 x 2.5"	600	3600.00
2.	PP bags	30 kg	-	150 Rs.	4500.00
3.	Rope	20 kg	-	90 Rs.	1800.00
4.	Straw	1 Trolly	-		3500.00
					36400/Rs.

b) Raw Material

Sl.No.	Material	Quantity	Rate	Amount
1.	Straw	20 Quintal	600 Rs./Quintal	12000.00
2.	Spawn	200 kg	150/per kg	30000.00
3.	PP bag and Rubber band	40 kg	150/per kg	6000.00

4.	Electricity	-	-	7560.00
5.	Others	-	-	2400.00
				57960/Rs.

c)

Sl.No.	Main Expenses	Total Amount	Rate	Amount
1.	Expenses on Bamboo	22600.00	20 per cent	4520.00
2.	Interest			
	* Room	35150.00	12 per cent 6 months	2109.00
	* Raw Material	57960.00	12 per cent 6 months	11592.00
	* Salary	32000.00	12 per cent 6 months	1920.00
				20141.00/Rs.

d) Total Expenses

Sl.No.	Material	Total Expanses
1.	Raw Material	57960 + 8000
2.	Labour	32000
3.	Interest	20141

e) Profit

Total Production	1700 kg
Rate	200/kg
Cost of Prod.	340000/Rs.
Total Profit	2,40,026 /Rs.

ANSWERS THE FOLLOWING

Q.1. When and where cultivation of Oyster mushroom takes place ?

Q.2. Mention the life cycle of *Pleurotus.*

Q.3. Discuss about the cultivation technology of *Pleurotus.*

Q. 4. Mention the Economic Importance of *Pleurotus* mushroom.

Q.5. What is the nutrient value of *Pleurotus* ?

Q.6. Discuss about the environmental conditions required for the growth of *Pleurotus.*

Q.7. Is cultivation of *Pleurotus* can be a good Entrepreneurship ?

Chapter 5

Cultivation of Milky Mushroom (*Calocybe* sp.)

Evolution of food culture have brought mushroom in the limelight. Its unique taste and nutritional value derived human understanding in civilized societies and ethnic food is becoming popular. One major problem of our planet is concern for food safety and security on one hand and huge loss from agricultural waste ~4,500 million tones/year. Microbial technology can help in large scale recycling of agrowaste in India.

Classification

Kingdom	–	Eumycota
Division	–	Basidiomycota
Class	–	Agaricomycetes
Order	–	Agaricales
Family	–	Lyophyllaceae
Genus	–	*Calocybe*

Various Species of *Calocybe*

Calocybe alneti, C. atropapillate, C. bipigmentata, C. cerina, C. chrysenteron, C. civilis, C. clusii, C. coniceps, C. constricta, C. cyanea, C. cyanella, C. cyanoceophala, C. eborina, C. fallax, C. gambosa, C. gangraenosa, C. georgii, C. indica, C. ionides, C. onychina and *C. rubra*.

Milky mushroom (*Calocybe indica*) is a tropical edible fungus and now it has greater demand due to robust, fleshy and milky white fruiting body. Milky mushroom grew well on uncomposted substrate under artificial indoor condition. Wide ranges of diverse cellulosic substrate are used for cultivation of milky mushroom. Paddy straw, wheat straw, soybean straw, sugarcane, bagasse, cotton waste and coconut coir pit are the common lignocellulosic substrates are used for cultivation of the mushroom. The name is derived from Ancient Great term Kalos (Pretty) and Cubos (head) (Nilson and Pearson, 1977). *Calocybe indica* can be easily grown in the temperature of 25-30°C. It has good biological efficiency and moderate protein content under optimum conditions. Its sporophore has long shelf life. The major advantage is that it can be best fitted in relay cropping when no other mushroom can be grown at higher temperature. *Calocybe indica* has a very good scope for further cultivation and it can replace the other tropical mushroom (Singh and Mishra 2005) (Figure 5.1).

Figure 5.1: Fruiting Body of *Calocybe* (Milky Mushroom).

Calocybe is a small genus of about 40 species of mushroom, which is edible and is cultivated in India. They are distributed in the tropical parts of the world (Shukla and Jaitely 2013). This mushroom is of Indian origin and it is widely distributed in Gangetic plains of West Bengal (Purkayastha and Chandra 1974). Due to its taste this mushroom has become third commercially grown mushroom in India, after button and oyster mushroom. Different strains of *Calocycbe* has been reported which show much diversity in their field depending upon the substrate used. First attempts on the induction of fruit bodies of *Calocybe indica* in culture was made by Purkayastha and Chadha in 1976. The mushroom species is suitable for hot and humid environment and can be cultivated indoor in high temperature and high humidity area. It grows well at a temperature range of 26-38°C and relative humidity more than 80 per cent. The first time milky mushroom was cultivated successfully in North Bengal in summer time (June-August) when the temperature and humidity is high which is suitable for milky mushroom cultivation. The cultivation process resembles that of oyster mushrooms but for the additional process of casing. The mushroom can be harvested after 30-34 days after spawning the advantage of this mushroom over other mushrooms are easy method of cultivation less investment, very attractive fruiting body pleasing milky white colour, long shelf life, more nutritious and less time to grow (Bokaria *et al.,* 2014). In rural areas, agrowaste and organic materials are available in plenty amount, which often end up in waste thus contributing to pollution control. Recycling is the need of the times and mushroom is an ideal candidate to harness its potentialities.

Advantages of *Calocybe*

1. It have antibiotic, antitumor and anticancer properties.
2. It helps in regulating diabetes, lowering cholesterol level and have strong antioxidant properties.
3. It is a good source of catalase, superoxide dismutase, Vitamin E, polyphenols, carotenoids. These compounds protects the human body against several chronic and degenerative disease.

Nutritional quality of mushroom is influenced by the substrate used, organic supplementation and other additive effect.

Know Yourself

Q.1. Mention the nutritional value of milky mushroom.

Q.2. How milky mushroom is different from oysters ?

Q.3. On which substrates it can be grown ?

Steps for the Cultivation of *Calocybe*

Mushroom strains, production of spawn, mushroom bag or bed preparation and incubation.

1. Cultivation chamber
2. Casing
3. Harvesting

Substrate and Substrate Preparation

Milky mushrooms can be grown on wide variety of substrates like oyster mushrooms. It can be grown on substrates containing cellulose, lignin and hemicelluloses substrate that to be used for growth should be fresh and dry. Substrates exposed to rain or harvested premature are prone to various moulds which may result in failure of the crop. It can be grown on straw of paddy, wheat, ragi, maize, bajra/cotton stalks and leaves, sugarcane bagasse, cotton and jute wastes, dehulled maize cobs, tea/coffee, waste *etc.* As in India wheat/paddy straw is easily available in abundance can be used (Figure 5.2).

Figure 5.2: Substrate for the preparation of compost.

Straw is chopped in small pieces (2-4 cm) size and soaked in fresh hot water for 6-8 hours. This period can be reduced when pasteurization is to be done by steam. Main purpose of soaking is to saturate the substrate with water. It is easier to soak, straw is filled in gunny bags and dipped in water. The substrate can be treated in various ways as follows:

Hot Water Treatment

Water is boiled in wide mouth container and chopped straw filled in gunybag is submerged in hot water for 40 minutes at 80-90°C to achieve pasteurization. This is very popular method particularly among small growers (Figure 5.3).

Figure 5.3: Hot water treatment.

Steam Pasteurization

Wet straw is filled inside insulated room either in perforated shelves or in wooden trays. Steam is released under pressure from a boiler and temperature in side substrate is raised to 65°C and maintained for 5-6 hours. Air inside the room should be circulated to have uniform temperature in the substrate (Figure 5.4).

Figure 5.4: Steam treatment.

Chemical Sterilization Technique

Ninety litres of water is taken in a rust proof drum (Galvanized) of G.I. tab of 200 litres capacity. Ten to twelve kg of wheat straw is slowly steeped in water. In another plastic bucket, Bavistin 7.5 and 125 ml formaldehycle (37-40 per cent) is dissolved and slowly poured on the already soaked wheat straw. Straw is pressed and covered with a polythene sheet. After 15 to 18 hr. the straw is taken out and excess water drained out. One can use a larger container or cemented tank of 1000-2000 litres for soaking more straw. Calculated amount of chemical can be added in the wetted straw at 0.1 per cent conc. It can be done by another method also. The straw filled in a bag can be soaked in water containing chemical solution. This can be a easier method, but this method is to be avoided due to use of chemicals.

Autoclaving

Substrate is filled in polypropylene bags (35 x 45 cm) holding 2-3 kg wet substrate and sterilized at 15 lb pressure. Once sterilization is over, straw is shifted to spawning room for cooling, filling in bag and spawning.

Spawning

This process can be done in bag or in beds.

Good quality spawn is used for the inoculation, close of 4-5 per cent of wet substrate is used, layer method of spawning is preferred. After spawning bags are shifted to spawn running room and kept in dark where temperature 25-35°C and relative humidity above 80 per cent are maintained. It takes about 20 days when substrate is fully colonized and bags are ready for casing (Figure 5.5).

Figure 5.5: Mycelium on the surface.

Mushroom Bed Preparation and Incubation

Black polythene bags of 60 x 30 cm size and 100 gauge thickness can be used and cylindrical mushroom beds can be prepared following layer methods of spawning. Beds can be made in which each bed contained 1.5 kg of paddy straw. A moisture content of about 60 per cent is required of the wetted substrate before the spawning. Wheat straw is spawned at the rate of 5 per cent of *Calocybe* spawn according to it's wet weight. Spawning can be done by layering method or by thorough method (Baskaran *et al.,* 1978). The beds will be incubated in semi dark room spawn running rooms at about 30 to 32°C. After 12-14 days beds will be fully colonized by the vegetative mycelium of mushroom fungus (Figures 5.6).

Figure 5.6: Bed preparation.

Casing

The upper surface of each bag/bed will be opened and a layer of casing soil will be applied on it upto the height of 2-3 cm. The beds will be regularly and uniformly sprayed with water to keep the surface of substrate moist. Casing provides physical support, moisture and allows gases to escape from the substrate casing materials (soil 75 per cent + sand 25 per cent) with pH adjusted to 7.8-7.9 with chalk powder is sterilized in autoclave at 15±lb pressure for one hour or chemically treated with formaldehyde solution 4 per cent about a week in advance of casing. Solution should be enough to saturate the soil with this certain amount of coco pit and cow dung manure will be added in casing soil. It is covered with polythene sheet to avoid escape of chemicals and at a interval of 2 days soil is turned so that at the time of casing soil is free from formalin fumes. Bags to be made uniform by ruffling top surface of the substrate and sprayed with solution of carbendazin (0.1 per cent) + formaldehyde (0.5 per cent). Casing material is spread in uniform layer of 2-3 cm in thickness and sprayed with a solution of carbendazin and formaldehyde to saturation level, temperature approximately 30-35°C and RH about 80-90 per cent will be maintained (Figure 5.7).

Figure 5.7: Preparation of Casing soil.

Fruiting

It takes about 10 days for mycelium to reach up to the casing layer, when fresh air is introduced while maintaining temp. 30-35°C and RH 80-90 per cent. Blue light should be provided for 10-12 hrs. daily. This results in the initiation of fruiting bodies with in 3-5 days in the form of needle shape which mature in about a week (Figures 5.8a-d). Mushrooms fruiting bodies of about 7-8 cm diameter can be harvested by twisting, clean and can be packed in perforated polythene/PP bags for marketing. Mushrooms can also be wrapped in Klin film for longer storage. From one bed of milky mushrooms, twice or thrice mushrooms can be harvested in 9 ± 3 days interval. No remnants of harvested basidiophores will be allowed to remain in the substrate. After such harvest, the casing soil will be slightly ruffuled. Finally the spent mushroom substrates will be collected and processed further for use as soil conditioner (Figure 5.8 a-d), 5.9.

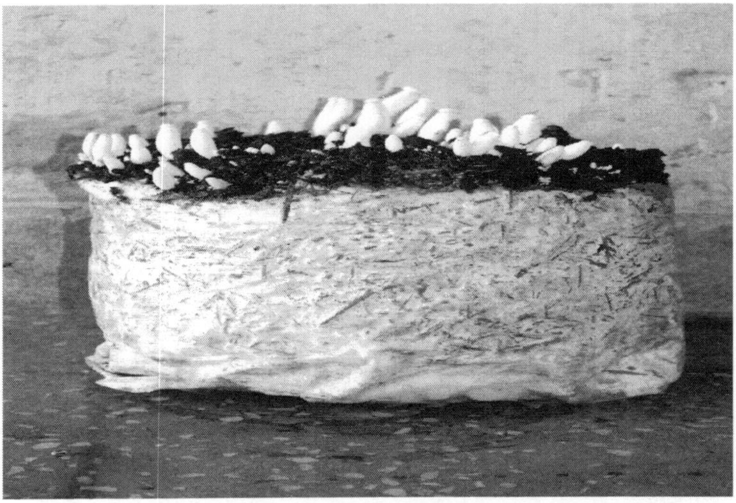

Figure 5.8a: Pinning of fruiting body on the compost.

Figure 5.8b Pinning stage

Figure 5.8c Young fruiting body

Figure 5.8d Mature fruiting body

Figure 5.9 Harvested fruiting body

Crop Management

Substrate is the major source of fungi/bacteria that's why substrate should be chopped and soaked in hot water properly, size of straw must be proper.

Room in which bag filling is to be done must be sterilized by 1 per cent formaldehyde spray twice a week.

For large scale cultivation Hepha filter can be used for air circulation in the room.

Room can be protected from flies and rats by placing nylon net or iron net on windows.

During casing if any patch of mould appears it can be treated at an early stage.

During rainy season watering should be done at controlled rate.

Cultivation Chamber

Cultivation chamber can be constructed in a shaded place, chamber lined with sky blue coloured light, high density polythene sheet coated with paddy straw as roofing materials. Beds after casing will be incubated over racks in the chamber. The side walls will be lined with paddy straw and plastic sheet. Sufficient ventilation can be provided from windows present in the chamber. The inside temperature ranges from 30 to 35°C and relative humidity is to be 85 per cent or can be more than that. During day time approximately fluorescence light must be available at least for 5 hrs. inside the chamber.

Preparation of Casing soil

Garden soil and sand will be sieved and dry. It is to mixed in the ratio of 4:1, autoclaved for 20 minutes. After 12 hrs. when the soil is cool it can be used for casing. Sterilized cocopeat and cow dung can be mixed in it. This soil to be spread uniformly over the spawn run beds. Regular spraying on the surface of the casing soil can be done to maintain moisture on the bed surface.

Brain Storming

Q.1. On which substrates Milky mushrooms can be grown?

Q.2. Define process of casing.

Q.3. Mention the constituents of casing layer.

Q.4. Draw the flow chart of Milky mushroom production steps.

Economics of Milky Mushroom Production

The cost of production depends upon the cost of raw material, yield/unit, production level and the whole sale price. The whole sale price in different parts of the country varies between 45-60/kg. The information given below is a model to work out cost of production for a medium size mushroom production unit keeping in view an average yield:

Sl.No.	*Item*	*Quantity*	*Cost*
1.	Paddy Straw	36500 kg	54,750.00
2.	Spawn	4600 kg	2,30,000.00
3.	PP bags for growing	500 gm	38,500.00
4.	PP bag for packing	500 gm	38,500.00
5.	Casing material	-	10,000.00
6.	Thread ball	-	2000.00
7.	Formaldeyde	50 lit.	3,750.00
8.	Bavistin	2.5 kg	2,000.00
9.	Melathion	2 lit.	1,600.00
10.	Bleaching Powder	50 kg	1,000.00
11.	Labour	3 Nos.	1,08,000.00
12.	Water	-	2,400.00
13.	Electricity	-	3,400.00
14.	Transport	-	10,000
15.	Rent of Shed	-	12,000.00
16.	Miscellaneous Expenditure		12,000.00
			5,20,900.00

Fresh Mushroom Production

Fresh Mushroom Produced	
Mushroom production	18250
Self Consumption	180
Wastage	180
Balance for Sale	17890
Realization at @ Rate 45/kg	8,05,500.00
Net Annual Profit	8,05,500.00
	5,20,000.00
	2,85,500.00
Cost of Product	**Rs.530/kg**

Source: Report.

Disease of *Calocybe*

Main pathogens of *Calocybe* are bacterial and fungal pathogens. Bacterial pathogens *i.e. Pseudomonas* sp. and *Bacillus* sp. caused bacterial blotch. While fungal pathogens are *Dactylium dendroides* and *Penicillium* cause disease on sporohore. Several fungi interfere with spawn running of milky mushroom are *Trichoderma* sp., *Coprinus* sp., *Chaetomium indicum, Rhizoctonia solani, Curvularia* sp., *Sclerotium* sp. and *Aspergillus* sp. These pathogens inhibit the mycelial growth on medium and substrate as well. Maximum infection can be seen in the month of June, when the humidity is very high and fungus mainly reported at that time is *Rhizoctonia solani*.

Management

Treatment of husk with $CaCO_3$ can be helpful is sterilization.

Phytoextract of garlic, *Ocimum*, turmeric, neem and marigold at 5, 10 and 15 per cent concentration can inhibit the infection. Turmeric and garlic extracts were most inhibitory to bacterial pathogens.

General cleaning, use of good quality substrate maintenance of optimum conditions will help to get substantial yield and reduction in disease incidence (Krishikosh).

Bokaria *et al.* (2014) reported that paddy straw, maize stalk and sorghun stalk gave significant higher yields followed by sugarcane bagasse and groundnut hulls. Physical and chemical state of substrates largely decide their suitability for mushroom growing. Poor growth and low yield of *Calocybe indica* in substrates could be related to the variation among the substrates in the bulk density, water holding capacity and improper gaseous exchange especially of harmful volatile compounds.

Role of Growth Regulations

Several growth regulators like Indole acetic acid, Indole-3 butyric acid, Gibberellic acid (GA) and kinetin affects the yield and size of sporophore of milky mushroom. Pani (2011) reported that spraying of GA at 50 ppm at the time of pinhead formation, greatly influenced the size of sporophore and yield of white milky mushroom.

Since this mushroom is morphologically similar to *Agaricus bisporus* it has been quite popular in Southern Indian States and slowly getting popular in other countries (China, Malaysia and Singapore). Small scale mushroom growers prefer to grow this tropical mushroom due to the following reasons:

1. Ideally suited to warm humid climate (30-38°C, 80 per cent to 85 per cent humidity).
2. Its longer shelf life without any refrigerator (can be stored upto 7 days at room temperature).
3. Retains fresh look and does not turn brown or dark black like that of *Agaricus*.
4. Lesser contamination due to competitor molds and insects during crop production under controlled conditions.

5. Infrastructure needed to grow this mushroom is affordable and cost of production is low, which means industrial production could be attractive.
6. Life cycle is short (7-8 week) and good biological efficiency of (140 per cent). Methods have been developed to increase its production in low cost (Krishnamoorthy and Balan, 2015).

Medically mushrooms are known to be abundant source of nutraceuticals which could decrease/reverse the progression of several disease. One such disease is diabetes mellitus, which in other wise characterized as hyperglycemia associated with insulin deficiency. Complication of this disease includes hypertension, artherosclerosis, microcirculators disorder and changes in large and small blood vessels (Sharma *et al.,* 2013).

Mushrooms are known to produce a wide range of volatile and flavour compounds with a distinctive profiles that very according to species, variety and sometimes cultural conditions (Rapior *et al.,* 1997). The flavor profile also changes when mushrooms are dried primarily due to the high level of oxidation. Usually these compounds are extracted using a combination of polar and non-polar solvents including water and dietyl-ether. Most of the flavour compounds include alcohols-aldehydes, ketones and oxides (Chandravandana *et al.,* 2005).

Answer the Following

Q.1. Discuss the cultivation process of Milky mushroom.

Q.2. How the cultivation of milky mushroom different from other mushrooms?

Q.3. Mention the nutrient contents of Milky mushroom.

Q.4. What are the medicinal value of Milky mushroom?

Q.5. How casing can be done?

Q.6. Mention some important species of *Calocybe*.

Chapter 6

Cultivation of Lingzhi (*Ganoderma* sp.)

Ganoderma is a genus of wood decaying polypore fungi, which is of great economic importance. It belongs to the:

Classification

Kingdom	–	Eumycota
Division	–	Basidiomycota
Class	–	Hymenomycetes
Family	–	Ganodermataceae
Order	–	Polyporales
Genus	–	*Ganoderma*

It is known as Reishi in Japanese, Lingzhi in Chinese and *Ganoderma* in English. More than 250 *Ganoderma* species have been described worldwide (Kaliyaperumal, 2013). In China, Lingzhi is considered as a herb of longevity. It has been commercially cultivated for the preparation of health products. In the later 1960s 2-3 species of *Ganoderma* were used to cultivate *i.e. G. lucidum, G. japonicum* and *G. capense*. Their incubation time was very long, for harvesting it took 2-3 years. In late 1980, new and improved cultivation method, short wood log have been gradually developed and employed for production of *Ganoderma*. With the recognition of medicinal properties of *Ganoderma,* commercial interest in *Ganoderma* products sharply increased during 1980-2000, not only in Asia but also in North America and Europe (Chen, 1999). Various techniques and different media have been utilized for *Ganoderma* production. Now a day's almost all *Ganoderma* producers prefer to adopt short wood-logs cultivation for production of *Ganoderma* fruiting body. For time saving, fermentation technique for production of *Ganoderma* mycelia, artificial cultivation has been started. *Ganoderma* gradually spread from China, Japan and the United States to all over the world.

Figure 6.1: Fruiting body of *Ganoderma*

Brain Storming

Q.1. Mention the common name of *Ganoderma* in various languages.

Q.2. Discuss the medicinal value of *Ganoderma*.

Q.3. Is *Ganoderma* fruiting body is soft or hard in texture?

Q.4. Origin place of *Ganoderma*.

Growing Conditions of *Ganoderma*

Ganoderma production goes under several stages: mycelium, primordium, young and mature fruiting bodies, each stage has a unique set of requirements in nutritional factors and growth parameters. The nutritional factors such as carbon and nitrogen source and growth parameters such as light, oxygen, temperature and relative humidity required at different development stages.

Nutritional Factors

a) Carbon Source

A carbon source contribute to the growth of *Ganoderma*. It supplies energy for growth and development of mycelia and fruiting bodies. The carbon sources are mainly organic, such as sugars, starch, cellulose, hemicellulose and lignin. The carbon sources commonly used are, sucrose and glucose for the cultivation of mycelia. For fruiting bodies carbon sources are hard wood sawdust, cotton seed hull, rice straw, wheat straw, corncob powder and other agricultural by products.

b) Nitrogen Sources

Nitrogen is the major component of protein. Without a nitrogen source proteins cannot be synthesized and mycelia cannot grow. *Ganoderma* can directly absorb amino acid, ammonia, urea and other small molecular compounds containing nitrogen. Yeast powder and peptone is added as a source of nitrogen in the media. Wheat bran, rice bran, Ammonium sulfate, Urea, corn powder can be used as a source of nitrogen during fruiting bodies formation.

Appropriate ratio of carbon to nitrogen is required for the growth and development of *Ganoderma*. The optimum ratio of carbon to nitrogen is 20:1, for sporulation of *G. lucidum* ratio is 30-40:1 (Zhou and Deng, 1996).

c) Inorganic Salts

Inorganic elements for the growth of *Ganoderma* include Calcium, Magnesium, Sodium, Phosphorus, Sulphur and Zinc. The most important among these are phosphorus, potassium and magnesium at the concentration of 100-150 mg/L. Other elements are provided in the form of Calcium sulfate (gypsum), Potassium-phosphate monobasic, and Magnesium sulfate in the media. Calcium sulfate is added at a dosage of approximately 1 per cent and is capable of adjusting the pH value of the substrate.

d) Growth Factors

For the growth and development of *Ganoderma* some growth factors are essential substances. That include B_1, B_2, B_6 and biotin, these compounds are required in little amount approximately 10 mg/L during the growth and development.

e) Moisture Content

Water is an important condition and component for growth. Cultivation of *Ganoderma* requires proper moisture content of the solid state substrate at approximately 60-65 per cent. For the growth of mycelia moisture content is required, sufficient amount of oxygen can be applied by it. Moisture content of the substrate can be increased to 70 per cent, when bagasse used as a substrate due to its large and loose space between material.

Environmental Factors

Temperature

Temperature is a necessary condition for enzymatic reactions for the development of the mycelia and fruiting bodies. *Ganoderma* mycelia can grow at the temperature range of 20-35°C, for spawn running temperature required is 30-32 degree centigrade, while for the fruiting body formation, temperature requirement with in the range of 24-28°C. At low temperature of 20°C, the fruiting body becomes yellow in colour and rigid, it dies at the temperature above the 35°C. If there will be great variation in temperature during growth period, the fruiting body will grow abnormal.

Moisture Requirement

During the run of spawn, the relative humidity must be maintained with in the range of 60-65 per cent. Moisture control is an important strategy for reducing fungal or bacterial growth in the indoor environment. During the fruiting body development period, the relative humidity with in the mushroom house can be increased to a level of 85-95 per cent.

Oxygen Level

Oxygen is one of the most important element required for the growth of *Ganoderma,* it is a high temperature aerobic fungi. To absorb oxygen and release of carbon dioxide during the whole growing period is necessary. The concentration of carbon-dioxide affects the growth of fruiting body. If the concentration of carbon dioxide is higher than 0.1 per cent the fruiting body will be abnormal in shape like the antlers. When carbon dioxide concentration is lower than 0.1 per cent then fruiting body becomes large thick and rounded cap with a short stipe is formed.

Answer the Following

1. **Mention some growth factors required for the cultivation of *Ganoderma*.**
2. **Which compounds can be the source of Carbon and Nitrogen during cultivation of *Ganoderma*?**

Light Condition

Ganoderma is very sensitive to light. Its growth is affected by light and dark conditions. Mycelial growth of the mycelium is inhibited by light, but during primordium formation weak light is very helpful, later on pileus formation takes place in the presence of light. Spawn generally run in the compost in absence of light, brief exposure of light triggers primordial formation and later on cap formation. Light intensity of 20-100 lux is favorable for antler formation, 3000-10,000 lux range of light can help in stipe and cap formation (Zhou, 2017).

According to (Zhou and Lin, 1999) optimum light level of 15,000-50,000 lux is required for the formation of stipes and caps.

Know Yourself

1. **How fruiting in *Ganoderma* is related to light intensity and carbon dioxide?**
2. **How anteler formation takes place in *Ganoderma?***

Cultivating Patterns

Like other edible mushrooms, the cultivation pattern of *Ganoderma* is same and can be divided into two major patterns

Solid State Medium

It involves the production of spores, mycelia, biomass and fruiting body formation (Figure 6.20).

Figure 6.2: Fruiting body on solid substrate

Liquid State Medium

Production of mycelial biomass formation takes place.

Production of *Ganoderma* can be divided into two steps.

1. First step involves the preparation of the culture, stock culture, mother spawn.
2. Second step involves the preparation of growth substrates for mushroom cultivation. Short wood log and synthetic saw dust bag and bottle method can be used for the fruiting body formation.

Wood Log Cultivation

In wood log cultivation technology, tree wood cut to length wise for artificial cultivation of mushroom. It can be divided into three types: Sterilizing the short wood log, wood log and stump cultivation. Short wood log cultivation method is advantageous as growth cycle is short, high yield and quality of fruiting body is good.

Substitute Cultivation

Substitute cultivation technology is by using the sawdust of hardwood, cottonseed husks or agricultural by products to cultivate mushroom (Triatana *et al.,* 1991). For the mycelial growth, fruiting body yield, various biologically active components are influenced by the substrate constituents and conditions. Rich raw materials can be utilized and divided into main ingredients and additives. Main ingredients are corncobs, sawdust, cottonseed and bagasse, additives include wheat bran, rice bran, corn flour and soyabean powder. Substitute cultivation can be done in bottles or bags. Bag cultivation has advantage over other methods, as moisture can be retained, fruiting bodies are bigger in size and transport of bags is handy, so it is widely used process.

Mostly *Ganoderma* is cultivated on solid-state media for production of fruiting bodies to obtain polysaccharides, triterpenoids, proteins and other metabolites used for the production of drugs for health products. Growing of *Ganoderma* is a long term process that requires one and several months for the fruiting bodies to appear, depending on species and substrate (Kites and Liu, 2000). *Ganoderma* fruiting bodies and its spore powders have long been used as a folk remedy for promotion of health and longevity, but they are not suitable for direct use as food or pharmaceutical ingredients, because they are hard and bitter to taste and harder for the body to absorb (Oru *et al.,* 2014). Development in modern technology and fermentation techniques can solve these problems.

Production of Mycelial Biomass

Modern fermentation techniques, provides potential to fast produce both mycelial biomass and metabolic products, in a shorter time with advantages of reduced space requirements and less chances of contamination. Fermentation can be divided into two types – liquid state fermentation and solid state fermentation.

	LSF		*SSF*
a)	Unitary product cost is high	a)	Unitary product cost is low
b)	Carbohydrates and nitrogen are major nutrients glucose, fructose, lactose, soyabean meal, cornflour, and organic nitrogen source is more suitable for the production	b)	As a substrate cheap raw material is used. Food and agricultural waste can be used for the substrate
c)	This method involves five stages: (1) Selection of strains(2) Preparation of medium for different culture phases, (3) Inoculation, (4) cultivation of selected strain (5) Harvest of *Ganoderma* mycelia to isolate products	c)	Purpose of SSF is to grow fungi into tight contact with insoluble substrate and to achieve the highest substrate conc. for fermentation
d)	Strain, temperature, pH, airflow and stirring rate play an important role in mycelial formation.pH value 5.5 is required for mycelial growth, 25-35°C temp. is favourable for its growth.	d)	Most appropriate for conversion of raw materials into value based products.

	LSF		*SSF*
e)	Little Magnesium sulfate and Potassium dihydrogen phosphate are added in media for better growth of mycelia.	e)	Advantage – low energy consumption, concentrated nutrient medium, high volumetric productivity concentrated end products.
		f)	Traditional and non-traditional substrates can be used efficiently rice, wheat, corn, soyabean *etc.*
		g)	Important factors are temp, pH, aeration, water activity and moisture, bed properties 40-60 per cent moisture.

Secondary metabolites produced are Ganodermic acids or polysaccharides, useful proteins and enzymes.

In the fermentation process, key factors for culturing and gaining effective products are level of carbon and nitrogen requirements, pH value, temperature, relative humidity, oxygen and nutritional components (Habijanic and Berovic, 2000)

Preparation of Substrate

Ganoderma is generally grown on sawdust, wood chips and hard wood logs. PP bags and glass bottles can be used to fill the substrate, for mycelium run and fruiting. Hardwood sawdust, cotton seed hull, crop straw and bagasse are generally used for the cultivation of *Ganoderma*. Cottonseed hull substrate gives the fruiting body of poor quality, while the saw dust substrate gives the quality of fruiting body good and surface is hard. Selection of raw material can be done properly (Figure 6.3-6.5) .

Figure 6.3: Wheat Bran

Figure 6.4: Saw dust

Figure 6.5: Mixing of substrate

Substrate formulation can be:

1. Sawdust 78 per cent, wheat bran 20 per cent, Sucrose 1 per cent and Gypsum 1 per cent.
2. Cottonseed hull 42 per cent, saw dust 42 per cent wheat bran 15 per cent and Gypsum 1 per cent.
3. Cotton seed hull 89 per cent, wheat bran 10 per cent, Gypsum 1 per cent.
4. Corncob 50 per cent, sawdust 35 per cent, wheat bran 15 per cent.
5. Corn cob 74.5 per cent, corn meal 24.5 per cent Gypsum 0.5 per cent plant ash 0.5 per cent.

These substrates can be modified according to the strains selected (Chen, 2014).

Think over it

Q.1. Mention the role of carbon and nitrogen ratio during fermentation process.

Q.2. How liquid state fermentation is different from solid state fermentation?

Q.3. Mention secondary metabolites obtained from the *Ganoderma*.

Packaging and Sterilization

While preparing the substrates, various ingredients should be weighed accurately and mixing should be thorough for cultivation in bag and bottle. Moisture content of the substrate should be approx. upto 55-65 per cent. The substrates are filled into the bottles or polypropylene bags. After that bottles/PP bags are plugged and covered with craft paper and autoclaved at 121°C for 2 hrs.

Preparation of Spawn and Inoculation

Spawn is the form of seed inoculums., It is used to inoculate the sterilized substrate after cooling. Spawn can be in liquid or solid form. Solid spawn can be saw dust based or grain based. Liquid mycelial spawn can be grown in potato dextrose broth or other formulations (Zhou *et al.,* 2007). For inoculum three steps are involved:

1. Preparation of Stock culture
2. Mother Spawn
3. Planting Spawn

Stock culture is usually stored on PDA slants at 4°C and transferred after every 2 months. While preparing mother spawn, cultures are grown on PDA plates at 25-27°C for 7-10 days in the absence of light. About 1 cm diameter mycelium from the plate is used for the production of liquid culture. The mycelia is incubated at 25-30°C in 150 ml PDB medium at 150-200 rpm shaker. After 8-10 days, mycelia will grow in liquid medium, when the mycelia reach to maximum biomass it can be harvested to use as planting spawn (Wang *et al.,* 2011). For solid spawn similar method is used.

After sterilization and cooling of the substrate, inoculation with liquid/solid spawn is done in a clean and dry place to avoid uninvited contamination. In wood log cultivation, amount of inoculum will be according to size of the log. In bag cultivation method, inoculum is placed on the substrate at the top in centre of the bag by making temporary hole to bottom through substrate. If the moisture content is high in the bag which will be poured out, leaked bags should not be inoculated otherwise it will be a source of secondary infection. 30-50 days old spawn or seed is suitable for the cultivation of *Ganoderma* (Chen *et al.,* 2013; Hou and Liao, 2009). In liquid spawn, inoculum value is set up with in the range of 5-15 per cent of the liquid medium (Figure 6.6-6.14).

Figure 6.6: Sterlization of PP Bags

Figure 6.7: Filling of PP Bags with substrate

Figure 6.8: Filled PP Bags with susbstrate

Figure 6.9: Making of hole in filled PP Bags

Figure 6.10: PP Bags with hole

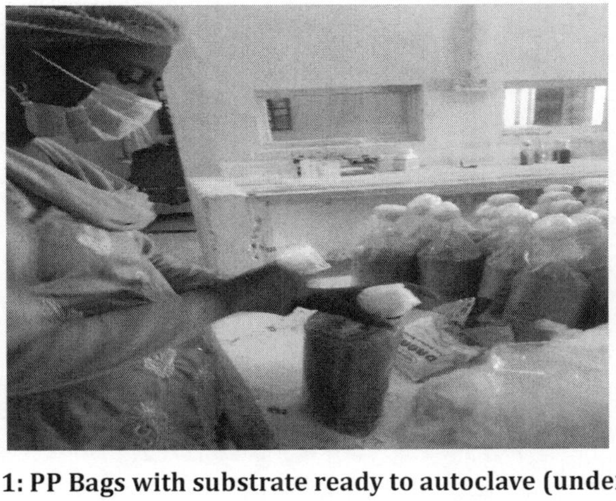

Figure 6.11: PP Bags with substrate ready to autoclave (under process)

Figure 6.12: PP Bags with substrate ready to autoclave

Figure 6.13: PP Bags with substrate in autoclave

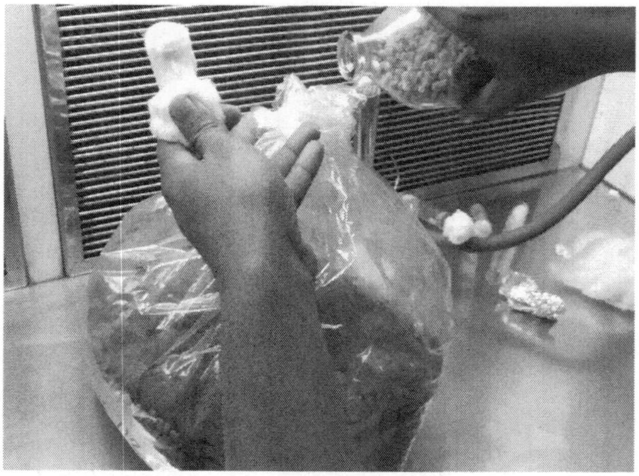

Figure 6.14: Inoculation of PP Bags with spawn

Spawn Running

After inoculation, bottles/bags are kept for spawn running at 26 ± 1°C in darkness. Relative humidity is adjusted to 60-70 per cent during the running of spawn. Proper temperature, ventilation and humidity should be maintained during this period. At 22°C mycelium will germinate and connected to each other with in a week. As the growth of mycelium takes place some droplets ooze out due to the respiration of mycelium. To avoid that we have to increase ventilation and reduce the humidity. Bags can be punctured by sterile needle or scissors so that water droplets can come out and aeration can happen. If temperature remains low during spawn running, it will take longer time in complete colonization of the substrate. After completion of spawn run in dark, diffused light (300/4x) will help in primordia formation. Strong light will inhibit the growth of mycelia and primordial formation. Brief exposure to light of mycelium triggers the primordial formation. In log cultivation it takes 60-70 days for primordial formation. Development process of fruiting bodies can be divided into three stages. Most crucial stage is pileus differentiation. During this stage, increase the ventilation, so that CO_2 concentration can be reduced along with high humidity and dim light (Chen, 2002). For the development of stalk, bags are transferred to fruiting room at temperature of 25 ± 1°C, 80-90 per cent relative humidity. Good quality fruiting body has dense texture and deep brown colour. Proper fruiting bodies will not differentiate at continuous high temperature of 35°C and low temp of 18°C. The management strategies during pileus differentiation involve some measures, such as water spray, light, aeration, prevention of insects diseases, drying. Variable temperature is not beneficial for the differentiation and development of fruiting body.

Ganoderma, fruiting body primordial differentiation happens at constant temperature conditions. The sarcomatous primordium can be formed at 12-14°C, stipe primordium at 15°C as the lowest temperature and pileus can be formed at 22°C as the critical minimum temperature. At the average temperature 18°C pileus formation takes place in 25 days, while at the temperature of 25°C it takes about 18 days. Moisture, lighting and ventilation affect the production of pileus. When the caps becomes completely red and white part from the margins disappear, than it is ready to harvest. Harvesting can be done by cutting the stipe with the help of knife, about 2 cm stipe should be left with pileus. After harvesting fruiting bodies can be sun dried or in hot air oven at 60°C, that will take 2-3 days. While drying, pileus part of the fruiting body should be placed towards down. During rainy or cloudy days, fruiting bodies can be dried by heat at 60°C. Improper drying lowers the quality of the products, lower side of the pileus can be contaminated by moulds and turned dark brown in colour. If conditions are maintained properly, second and third flushes can be harvested; although the yield is lower in these flushes. After 2 weeks, second flush can be harvested. If appropriate conditions and optimum management is practiced, 1 kg of dry substrate can produce 100-120 gm of dry fruiting body of *Ganoderma* (Figure 6.15-6.16).

Figure 6.15: Incubation of PP Bags with substrate and spawn

Figure 6.16: Fruing of *Ganoderma* in Bags

Brinstroming

1. **Mention various methods of *Ganoderma* cultivation.**
2. **How logs can be prepared for the spawning of *Ganoderma*?**
3. **Mention the environmental conditions required for cultivation of *Ganoderma*.**

Medicinal Value of *Ganoderma*

In developing countries, traditional medicines held an important status in health care system. *Ganoderma* has an important place among the traditional medicines. About 400 different bioactive compounds have been isolated and identified from the fruiting bodies, spores and mycelia of *Ganoderma*. Active compounds include polysaccharides, triterpenoids, nucleotides, sterols, steroids, fatty acids, proteins/peptides and other trace elements. These components have many pharmacological effects such as antitumor, antibacterial, antiviral, sleep promoting immunomodulation, antiaging, antiulcer properties (Sanodiya *et al.,* 2009). Fungal immunomodulatory protein **(FIP)** is also an important bioactive component, with immune regulating activity, which is the most promising active ingredient developed by modern biotechnologists (Li *et al.,* 2011a). Polysaccharides, were the first discovered ingredient from *Ganoderma*. α or β (1-3), (1-6) Glucans, heteropolysaccharides with different combinations of glucose, mannose, galactose, xylose, fructose as well as arabinose have been isolated from various species of *Ganoderma* (Nie *et al.,* 2013). **Ganopoly** is a Glucan β1-3 and β1-6 D-Glucans mainly consisting, of neutral polysaccharides of glucose units Ganoply are mediated by its immunomodulatory, anti angiogenic and cytotoxic effects. Ganoply affects immune cells and immune related cells including B-lymphocytes, T- lymphocytes, dendritic cells, macrophages and natural killer cells (Xu *et al.,* 2011).

Triterpenes are another type of most important biological active compounds in *Ganoderma*. More than 150 types of terpenoids have been isolated from the fruiting bodies, spores and mycelia that can be used in pharmacological activities. Triterpenes are anti-allergy, anti-oxidation, anti-hypertension, hepatoprotection, cholesterol reduction synthetase and so on. Lin *et al.* (2003) reported that fraction amount of triterpene could inhibit the growth of human hepatoma cells via suppressing protein kinase C and activating mitogen activated protein kinase and C-2 phase of cell cycle arrest. There are reports that number of triterpenes show significant anti HIV-1 protease activity (Zhu *et al.,* 1999).

Reishi mushrooms can enhance immune function through its effects on white blood cells. Which helps in fighting with the infection and cancer. Its in use for cancer prevention and treatment.

Ganoderma helps in decreasing anxiety and depression, as well as help in improving quality of life.

It helps in controlling blood sugar improve good cholesterol and antioxidants in body.

Ganoderma is considered a very safe and gentle tonic herb that can be tolerated by almost anyone.

Traditionally known as mushroom of Immortality for its extensive range of longevity promising benefits

It promote better night sleep, nourish and protect skin.

Elevate mood and happiness.

Fungal immunomodulatory proteins – a small molecule protein have been isolated from *G. lucidum, G. japonicum, G. microsporum, G. sinensis etc.* Various studies provides that it had antitumor activity both *in vivo* and *in vitro*. It will become a potential drugs for anti cancer activity with high efficiency, low toxicity, high specificity and conducting cancer chemical prophylaxes and antitumor therapies.

Economical Importance of *Ganoderma*

These days *Ganoderma* is widely accepted as a valuable health supplement and herbal medicine for the prevention and treatment of many diseases.

- It is known for its medicinal and nutritional value.
- In traditional methods it is used as an alternative of radiation therapy or chemotherapy.
- *Ganoderma* products are available in market as a nutritional supplements, Dietary supplements, functional foods and myco pharmaceuticals.

Ganoderma coffee is a drink mixture available in powder form. It typically contains instant coffee and the powdered extract of *Ganoderma lucidum*. Other ingredients such as sugar, non-dairy creamer, and herbs can be included in it. Some times people used *Ganoderma* coffee to reduce their actual coffee intake and boost their energy level. *Ganoderma* coffee can improve memory, fight fatigue, lower cholesterol, boost immunity, relieve stress, reverse aging process, increase energy stamina and stimulate circulation. Drinking *Ganoderma* coffee help to reduce risk of developing several chronic diseases *i.e.* diabetes, Parkinson's disease and heart disease. With all these positive side, it have side effects also *i.e.* it may cause dizziness, stomach upset and skin irritation, still research is going on *Ganoderma* and its health effects (Figure 6.17-6.19).

Figure 6.17: Chips of *Ganoderma*

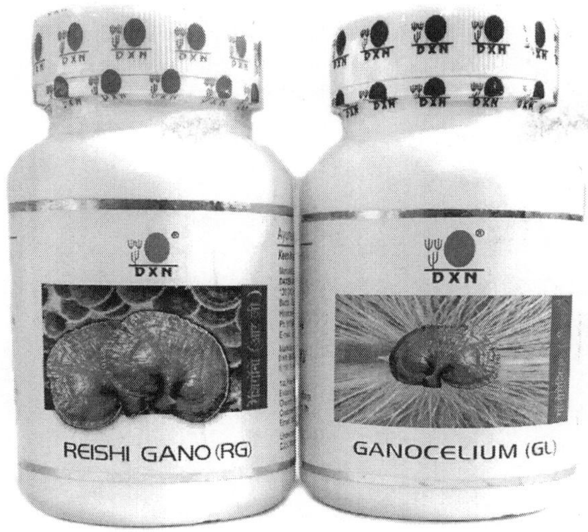

Figure 6.18: Commercially available tablets of *Ganoderma*

Figure 6.19: Mature Fruiting body of *Ganoderma*

Think over it

1. **Commercially available product of *Ganoderma*.**
2. **Mention the benefits of *Ganoderma* coffee.**

Other Components

Elemental analysis of log-cultivated fruit bodies of *G. lucidum* revealed Calcium, Potassium, phosphorus, Silica and Magnesium to be their main mineral compounds. Iron, Sodium Zinc, Copper, Maganese and Strontium were also detected in lower amounts, heavy metals *i.e.*, lead, cadmium and mercury are also present (Chen *et al.*, 1998). Freeze dried *Ganoderma* bodies collected from wild have mineral content of 10.2 per cent with Potassium, Calcium and Magnesium as the major component (Chiu *et al.*, 2000). Protein contents of dried *G. lucidum* was found to be around 7-8 per cent, which is lower than other mushrooms (Chang and Buswell, 1996). Bioactive proteins contribute to the medicinal properties of *G. lucidum i.e.* LZ-8 an immunosuppressive protein purified from mycelia (Van Der Hem *et al.*, 1995). Peptide (GLP) exhibit hepatoprotective and antioxidant activities and 15-KDA antifungal protein (Ganodermin), which is isolated from *G. lucidum* fruiting bodies (Wang and Wang 2006).

From the dried mushrooms carbohydrates and crude fiber content were examined and 26-28 per cent and 59 per cent was found respectively (Mau, Lin and Chen, 2001). Lectins were isolated from fruit-body and mycelium which have hemagglutinating activity on pronase-treated human erythrocytes (Thakur *et al.*, 2007). An enzyme metalloprotease has been isolated from *G. lucidum* which delays clotting time: ergosterol (vitamin D2) nucleosides and nucleotides (Paterson, 2006).

Brain Teaser

1. **Which bioactive compounds are present in *Ganoderma*?**
2. **Name the heavy metals found in *Ganoderma* fruiting bodies.**
3. **Which diseases can be cured by *Ganoderma* ?**
4. **Name the antifungal protein isolated from *Ganoderma* fruiting body.**

Disease of *Ganoderma*

During poor environmental conditions like temperature, humidity, and toxic chemicals, infection of pathogens takes place. Metabolic disorders and abnormal changes takes place in morphological and physiological features of *Ganoderma lucidum* fruiting bodies. Most common pathogens on *Ganoderma* takes place during its cultivation process are:

Brown rot disease caused by *Mycogone perniciosa,* takes place during cultivation process. During this infection, fruiting bodies grows abnormally with enlarged stipe and small cap, surface of fruiting body becomes covered by white mycelium, later on it changes to brown in colour and brown colour droplets oozes out.

Mildew disease is caused by *Trichoderma* which is a pathogenic fungi during cultivation process. This disease causes great yield loss as the fruiting body formation is affected. Mucus is exuded from the cap surface and fruiting body stops growing and becomes green in colour. As mycelium of *Trichoderma* secrets toxins, cause inhibition of mycelial growth on the substrate.

Aspergillus species also slow downs the growth of mycelium during cultivation, fruiting body formation is affected due to infection. *Neurospora* is a ascomycetous fungi, which grows very fast on the substrate of *Ganoderma i.e.*, log cultivation as well as compost-substrate cultivation. These diseases seriously affects the growth and yield of the fruiting bodies. Insects *i.e. Isopteran, Colcoptera, Lepidoptera* and *Collembolan* affects the fruiting bodies. Moths, beetles, termites, nematodes, slugs also affects the fruiting body formation.

Prevention of Disease

These disease can be controlled by physical, chemical and agricultural control methods:

Physical Control

The cultivation site must be sanitized properly for repeated cultivation to avoid the growth of insect eggs and fungal spores. Fly net, sunshade net can be used to avoid flies. UV and ozone light can be helpful in controlling pest at the inoculation apparatus and cultivation site.

Chemical agents are quite effective in control and preventing of disease and pest, during cultivation. Chemical can be harmful for the human health so it can be used only when it is necessary, chlorothalonil, thiram and prochloran 30 per cent can be used for spraying.

Substrate to be used should be sterilized completely. Environmental conditions must be hygienic, equipments must be treated with disinfectant, workers hands must be sanitized properly.

Answer the Following

1. **Discuss the process of cultivation of *Ganoderma*.**
2. **Mention the bioactive compounds present in *Ganoderma*.**
3. **Mention the medicinal properties of *Ganoderma*.**
4. **Mention the Economic importance of *Ganoderma*.**
5. **Discuss the diseases occurs on *Ganoderma* substrate during cultivation and how it can be controlled?**
6. **What challenges one can face during the cultivation of *Ganoderma*.**
7. **Discuss the method of bed preparation during inoculation process.**

Other Medicinal Mushrooms

Mushrooms are becoming very popular these days due to their nutrient value. They are known as nutraceuticals also, full of antioxidants and other chemical compounds. Most of mushrooms are growing naturally in nature and can be used in the cure of many diseases. Even some mushrooms are anti-cancerous in nature can be the alternate of chemotherapy, immunity booster.Here we are going to mention about some medicinal mushrooms *i.e., Cordyceps* and *Hericium* can be cultivated in laboratory, commercially people are growing these mushrooms. Chemical compounds present in them are isolated and used to make medicines.

Cordyceps militaris **– Kidajadi or jumbi mushroom**

Kingdom	–	Eumycota
Division	–	Ascomycota
Class	–	Sordariomycetes
Order	–	Hypocreales
Family	–	Cordycipitaceae
Genus	–	*Cordyceps*
Species	–	*militaris*

Figure 6.20: *Cordyceps*

1. *Cordyceps is* popularly known as kidajadi .
2. Commonly it is found in Himalayan regions of India and Nepal.
3. It provides immunity, stamina, energy and sleep quality.
4. Many active compounds are present in it.
5. Cordycepin is the most common compound present in it.
6. It increases, flow of blood, improves health of heart, enhances kidney function.
7. Cordycepin present in this mushroom is just like a magic drug, helps in opening of blockage of heart.
8. It can be an alternative of chemotherapy.
9. It is useful for lungs, heart, reproductive organ, brain and blood vessels.

Hericium coralloides- Lion's mane mushroom

Kingdom	–	Eumycota
Division	–	Basidiomycota
Class	–	Agaricomycetes
Order	–	Russales
Family	–	Hericiaceae
Genus	–	*Hericium*
Species	–	*coralloides*

Figure 6.21: *Hericium coralloides*

1. *Hericium coralloides* is commonly known as Lion's mane mushroom or tooth fungi.
2. It is an edible and medicinal mushroom
3. Lion's mane mushroom is native of North America, Asia and Europe.
4. It generally grows on hard woods, have spine like structure on it.
5. It grows in a single clump of having hard spines.
6. It can grow as a parasite on tree, so known as endophytic fungi.
7. It taste like lobster, so people in Europe prefer to eat it.
8. Bioactive compound present in it is Polysaccharides, Beta Glucan, hericenones and erinacines.

Inonotus obliquus - Chaga mushroom

Kingdom	–	Eumycota
Division	–	Basidiomycota
Class	–	Agaricomycetes
Order:	–	Hymenochaetales
Family	–	Hymenochaetaceae
Genus	–	*Inonotus*
Species	–	*obliquus*

Figure 6.22: Fruiting body of *Inonotus obliquus*

1. ***Inonotus obliquus*** is known as chaga mushroom.
2. Grows in Northern Hemisphere, in cold climate.
3. As a parasite it grows on birch trees.
4. From outside it is black in colour but from inside it is orange in colour.
5. It is rich in antioxidants, due to that it may benefit health.
6. Chaga mushroom is available in the form of powder and herbal tea.
7. It helps in reducing bad cholesterol, reduces stress and support immune system.
8. It have melanins, polyphenols, flavans, polysaccharides and triterpenoides.

Answer the Following

Q.1. On the basis of uses mention various types of mushroom.

Q.2. Discuss about the classification of various medicinal mushrooms.

Q.3. Which phytochemicals are presents in the mushrooms?

Q.4. Mention the benefits of chaga and Lion's mane mushroom.

Q.5. Why *Cordyceps* is known as Kidajadi ?

Q.6. Which chemical compounds are present in *Ganoderma?*

Poisonous mushrooms

Most of the mushrooms produces secondary metabolites which can be poisonous, mind altering antibiotic or bioluminescent. Even though there are some deadly species, which can cause severe symptoms of illness. A number of most toxic species such as *Amanita biosporigera* seem like edible mushrooms. In various phases of growth, they can readily be confused with button mushroom, puff balls or other edibles. *Amanita phalloides* is accountable for the vast majority of the mushroom-ingestion related deaths of people and known as death cap. Other species of mushrooms belongs to Genus *Psilocybe* may create hallucinations and other results on the mind. Some mushrooms have harmful effects from ingestion of toxic substances present in them and that is known as **mycetism** or mycetismus. Symptoms of this effect varies from gastrointestinal problem to death. These toxins are secondary metabolites produced by the fungi. There is very thin demarcation line between the edible and the poisonous mushrooms.

Salient features of poisonous mushrooms

- ☆ Scales are present on the pileus part of the mushroom.
- ☆ Gills are white in color, a ring like structure is present on the stipe.
- ☆ At base, swollen structure is present known as volva.
- ☆ Poisonous mushrooms have bad odour.
- ☆ They are bitter in taste. Small amount of mushroom placed on tongue, gives burning sensation.
- ☆ Spore can be green, purple or white in color.
- ☆ There is absence of worms around them.

NOTE:- These are not hard and fast rule regarding the identification. Some exceptions are always there.

Amanita phalloides

Kingdom	–	Fungi
Division	–	Basidiomycota
Class	–	Agaricomycetes
Order	–	Agaricales
Family	–	Amanitaceae
Genus	–	*Amanita*
Species	–	*phalloides*

Figure 6.23: Fruiting body of *Amanita phalloides*

- *Amanita phalloides* is white or light yellowish in color.
- Scales are present on the upper surface of the pileus.
- It is commonly known as death cap, causes most dangerous mushroom poisoning.
- It has a bitter taste and smell is unpleasant.
- Primary symptoms of poisoning are dehydration, accelerated pulses, cramps in the legs and lowering of blood pressure. Indigestion also takes place.
- Amatoxins is the compound responsible for the mushroom poisoning in humans.

Amanita muscaria

- Most of the symptoms are common as that of *phalloides.*
- Symptoms generally occur after the consumption of these mushroom.
- Nervous excitement, shivering and tendency collapse are the main symptoms.

Figure 6.24: Fruiting body of *Amanita muscaria*

Amanita virosa

Figure 6.25: Fruiting body of *Amanita virosa*

- This species is also highly toxic in nature, also known as "Death Caps".
- Amatoxins and phallotoxins are also present in these mushrooms.
- On its consumption, the person feels diarrhoea, stomach pain and nausea initially. But at later on, it leads to kidney and liver damage.

***Coprinopsis atramentaria* (Inky cap mushrooms)**

Kingdom	–	Fungi
Division	–	Basidiomycota
Class	–	Agaricomycetes
Order	–	Agaricales
Family	–	Psathyrellaceae
Genus	–	*Coprinopsis*

Figure 6.26: Fruiting body of *Coprinopsis atramentaria*

- These are commonly known as Inky Cap Mushrooms.
- Commonly found in Europe and North America.
- It produces unpleasant smell.
- It causes poisoning when eaten along with alcohol or drinks. Otherwise, it is an edible mushroom.
- It causes reddening of the face, increased rate of heart beat, vomiting and diarrhoea.
- The active component is coprine.
- *Coprinus comatus* is an edible mushroom and it doesn't contain coprine.
- *Coprinus atramentaria* has gills that turn black and eventually liquify and creating an ink.
- It tends to grow in clusters from senescent systems around stumps. It has a grey to brownish grey cap that features a few tiny scales over the centre but lacks prominent scales.
- Odour and taste of these mushrooms are very different.
- The color print of spore is black in color.
- Shelf life is very less, approximately 24 hours.
- They are nutrient dense survival food in a pinch.

***Cortinarius orellanus* (Fool's webcap mushrooms)**

Kingdom	–	Fungi
Division	–	Basidiomycota
Class	–	Agaricomycetes
Order	–	Agaricales
Family	–	Cortinariaceae
Genus	–	*Cortinarius*

Figure 6.27: Fruiting body of *Cortinarius orellanus*

- Orellanin is the main toxin present in these mushrooms.
- It causes orellanus syndrome.
- It is nephrotoxic.
- It causes gastrointestinal complaints, back pain and renal failures.
- Orellanin undergoes photochemical degradation into toxic Orellanine.

Psilocybe cubensis (Magic Mushrooms)

Kingdom	–	Fungi
Division	–	Basidiomycota
Class	–	Agaricomycetes
Order	–	Agaricales
Family	–	Hymenogastraceae
Genus	–	*Psilocybe*

Figure 6.28: Fruiting body of *Psilocybe cubensis*

- These are commonly known as "Magic Mushrooms". It includes many members of *Psilocybe.*
- It appears in egyptian tomb paintings.
- The extraordinary fragment possibly depicts the brewing and consumption of mythical ritual drink some was found at the burial site in Northern Mangolia.
- It contains psilocybin, psilocin, beaocybin.
- As it contains psycoactive compounds.
- Symptoms are dizziness, hallucinations, headache, weird hillarity, bone ache, depression, vomiting and ataxia.

Answer the Following

Q.1. Name some Poisonous Mushroom.

Q.2. Name the chemicals/bio-active compound present in Poisonous Mushroom.

Q.3. Why Magic Mushroom are known as Magic Mushroom?

Q.4. What symptoms can be there after consuming Poisonous Mushroom?

Q.5. Why it is not advisable to eat Mushroom collected from woods?

Chapter 7

Cultivation of Paddy Straw Mushroom (*Volvariella* sp.)

Paddy straw mushroom (*Volvariella volvacea*), commonly known as the straw mushroom or the Chinese Mushroom belongs to family Pluteaceae of the Basidiomycetes. It is most popular in South East Asia. Several species of *Volvariella* have been grown for food *V. bombycina* and *V. diplasia* have been cultivated in India. *Volvariella diplasia* is white in colour, white *V. volvacea* is blackish, further *V. bombycina* differs from *V. volvacea* in terms of colour as well as habitat.

Classification of *Volvariella*

Kingdom	–	Eumycota
Divison	–	Basidiomycota
Class	–	Agaricomycetes
Order	–	Agaricales
Family	–	Pluteaceae
Genus	–	*Volvariella*

Volvariella is an edible mushroom cultivated in tropics and subtropics. First time it was cultivated in China in 1822. Initially, this mushroom was famous by the name of "Nanuha Mushroom" after the name of Nanhua Temple in North Guangdong Province in China. Buddhist monk started the cultivation of paddy straw mushroom for their use. About 300 years ago its cultivation became popular

(Chang, 1971). Around 1935 *Volvariella* was introduced into Malaysia, Philippines and other South Asian countries by Chinese (Baker, 1934). *Volariella* is also known as "Warm Mushroom" as it grows at higher temperature. It is a fast growing mushroom, under favourable conditions for growth with 4-5 week time its crop cycle can be completed. It can be cultivated in the temperature range of 28°C to 38°C and relative humidity of 57-60 per cent. Ideal temperature for spawn run is 30-35°C while for fruiting 28-32°C. For spawn run and fructification RH of 80-85 per cent is ideal (Figure 7.1).

Figure 7.1: Fruiting body of *Volvariella*

Volvariella as known as paddy straw mushroom, generally grows on paddy straw substrate. However, it can be cultivated on cotton waste, cotton ginning mill waste, cotton spinning mill waste, sugarcane bagasse and banana pseudo stem *etc.* Nutritionally this mushroom contains high level of protein (3.90 per cent on fresh weight basis), good amount of Iron, Calcium, Potassium and Vitamin B complex.

Advantage

Being a high temperature living mushroom, it can be cultivated in major part of India and other countries. It is considered as most tasty mushroom among the edible mushrooms. It has short cropping cycle and one crop cycle is completed in 20-25 days. Cultivation technology is very simple and can be cultivated outdoor and indoor.

Ask Yourself

1. **Why *Volvariella* commonly known as Nanuha Mushroom.**
2. **First time in which region it was grown?**
3. **What can be the substrate for *Volvariella* Mushroom ?**
4. **Which nutrients are present in *Volvariella* Mushroom ?**

Regions in the Country where it is Cultivated

Orissa is the leading state, where it is cultivated mainly out door under shadow of coconut, cashew, mango, banana and teak trees. It is popular in states of West Bengal, Chhattisgarh, Jharkhand, Bihar, Andhra Pradesh, Tamil Nadu *etc.*

Life Cycle of *Volvariella*

Paddy straw mushroom has a distinction from other mushrooms. Being homothallic species the individual uninucleate, haploid, self fertile spores germinate to produce mycelia and completes the life cycle without the need of a mating type factor. Clamp connections are absent in *Volvariella* species. Hyphal cells are multinucleate, clamp connections are absent and basidiospore receives only one nucleus each, following meiosis. Wide variation exists in growth rate and other characteristics of single spore mycelia. It is still difficult to call this a homothallic mushroom as research is going on.

Fruiting body of the *Volvariella* is divided into six different development stages *i.e.* Pinhead, tiny button, button, egg, elongation and mature stage. Each stage has its own morphology and anatomy (Figure 7.2).

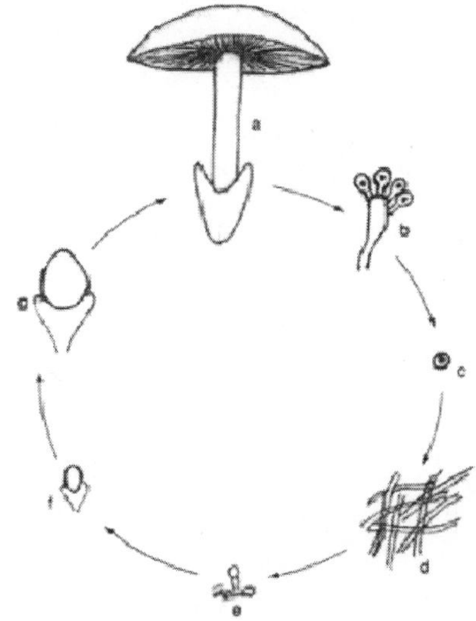

Figure 7.2: life cycle of *Volvariella*

Pinhead Stage

During Pinhead stage, mushroom is of the size of Pinhead, veil is spotlessely white in colour. In vertical section Pileus and stipe cannot be seen. While structure looks like a knot of hyphal cells.

Tiny Button

Both the tiny button and pinhead stages are formed from interwoven hyphae. At this stage, the top of the veil is brown, while rest is white. It is round in shape, if vertical cut at button stage is made, lamellae can be seen as a narrow band on the lower surface of pileus.

Button Stage

This stage of paddy straw mushroom is short, while the stipe attains the full length. Whole structure is wrapped in a coat, which is called as the Universal veil. Inside the veil, closed pileus exists. Stipe is not visible but when it is cut longitudinally then stipe can be seen.

Egg Stage

At egg stage, pileus comes out of the veil, Veil remains attached as a *volva*. The stipe is again not visible at this stage. The lamellae of this stage do not bear basidiospores. The size of the pileus remains very small up to this stage.

Mature Stage

During this stage, structure is divided into three region:

1. Pileus or cap
2. Stipe or stalk
3. Volva or cup

Pileus is circular structure of 6-12 cm in diameter and is connected in the centre with a stipe. Surface of the pileus is smooth and dark grey at centre and light grey at the margin. Lamella is present at the lower surface of pileus and number varies from 280-380. The lamellae vary in size from full size to one quarter size of the pileus. Each lamellae is formed of three layers of interwoven hyphae. The outermost layer is called hymenium, and it forms club shaped basidia and cystedia. Basidiospores are present on basidia, four basidiospores are present on each basidium. Basidiospores vary in shape, egg shaped, spherical or ellipsoidal. The colour of basidiospores again varies and it may be light yellow, pink or dark brown. The length of stipe depends upon the size of the pileus, and it is usually about 3 to 8 cm in length and 0.5 to 1.5 cm in diameter. It is white, fleshy and without any annulus. At the base of stipe remains the volva, which is a thin sheet of interwoven hyphae around the bulbous base of the stipe. The volva is fleshy, white and cup shaped with irregular margins. The base of volva bear's rhizomorphs which absorb the nutrition from the substrate.

A potential high yielding strain of *Volvariella* is apparent based on luscious morphological characters, myceial growth rate, type of mycelia, aerial hyphae formation and chlamydospore production (Chang and Yau, 1971). The asexual chlamydospore is a thick walled vegetative cell, normally formed at hyphal tips and are visually represented as a pink coloured dots in culture. Factors such as pH, low temperature and oxygen limitation are important inducers of morphogenetic pathway required for chlamydospore formation (Kumar *et al.*, 2016).

Methods of Cultivation of Mushroom

The cultivation of paddy straw mushroom can be done in a thatched house and also under the shade of a tree. Fresh, diseases free paddy straw is the ideal substrate. Ten-fifteen kg paddy straw is necessary for preparing one bed. These days it is cultivated inside plastic film house to maintain the temperature of around 25-35°C and relative humidity 75-80 per cent (Figure 7.3).

Figure 7.3: Different systems of cultivation

Paddy Straw Bundle Method

- ☆ Prepare a raised platform of about 1m in length and.75m in breadth with a wood planks and keep it over a support by arranging bricks on all four corners.
- ☆ The paddy straw is bundled into the weigh about 2- 5 kg each.
- ☆ Soak the straw bundles in water for 12-18 hr.
- ☆ Then bundles are taken out and drain the excess water.
- ☆ Place the bundles over the platform with their butt end on one side.
- ☆ Build the second layer by placing the butt end towards the other direction (these 8 bundles make one layer of bed).
- ☆ Place small quantity of spawn 0.8-1 cm inside the margin at an interval of 10-15 cm all along the periphery.

- Apply a layer of Dhal powder after placing the spawn.
- Place the straw bundles at right angles to the previous layer in criss cross fashion to make the third layer.
- Place the straw bundles with opposite butts end to make the fourth layer.
- Spawn this layer as stated above.
- Place another layer of straw bundles over this and do not apply spawn.
- Pressure the bed to make it as compact as possible and cover it with a transparent polythene sheet.
- Keep the beds undisturbed for a few days.
- Success of cultivation depends upon the temperature and moisture in the bed. The optimum temperature of 30-35°C is necessary for developing buttons.
- Mushrooms start appearing from all sides in 6-10 days as tiny buttons, which can be harvested in another 4-5 days. The harvesting can be done at the button stage only, as opened sporocarp is more fibrous to handle. 1-2 kg of mushroom can be harvested from 10 kg substrate (Figure 7.4-7.6).

Figure 7.4: Bundle preparation

Figure 7.5: Bed preparation and spawning

Figure 7.6: Different stages of growth of *Volvariella*

- Make the straw into twist of about 5-8 m long and 5-10 cm diameter.
- Immerse twists in water for 12 hr.
- Take out the straw and drain the excess water.
- Place them length wise at a platform in a zigzag manner.
- Place a second layer over this in a opposite direction.
- Sprinkle the coarsely powdered dhal and place small bits of spawn all along the periphery.
- Build another layer and spawn the layer.
- Build 4-5 layers and spawn as usual.
- Compact the bed by pressing and cover it with a polythene sheet.

Fruiting bodies will come out with in 6-10 days from all sides.

Cultivation of *Volvariella* can be described in the Form of Outdoor Cultivation and Indoor Cultivation

Traditionally mushroom cultivated on beds made of uncomposted and unpasteurized bundles of banana leaves or rice straw, tied at both ends and laid one on top of the other to form a bed. Mushrooms beds which are exposed directly to the sun are covered with plastic sheets. The yield of mushroom from beds is unstable and irregular. Bed is subjected to microbial contamination and fluctuating environmental conditions.

Indoor cultivation method has three salient features:

Composting, pasteurization and cultivation inside a mushroom house using stacks of open wooden crates. Composting is an important process that allows the microbial decomposers to reduce the tensile strength of the rice straw. This also prepares the straw for easy colonization by the mycelium of the paddy straw mushroom. Pasteurization, on the other hand, is a critical process that eliminates undesirable microorganisms which may compete with the paddy straw mushroom during production. The cultivation of mushrooms in growing houses reduces the sudden fluctuations of temperature and relative humidity associated with outdoor production which can be hazardous to growth and fruiting body production (Reyes, 2000).

A good substrate provide good anchorage, good aeration and have water holding capacity and also contains the necessary nutrients, released from the raw compost materials during fermentation. When mushroom spawn is planted, it can spread throughout the compost in short time with little competition. *Volvariella* mycelium grows both on composted and non-composted substrates.

1. Non Pasteurized Portable Outdoor or Indoor beds

This process can be done in wooden boxes:

Material

Wooden box, paddy straw/banana leaves, plastic sheets/house for mushroom, Frances of 60 x 45 x 20 cm size open from all sides.

Bedding material are cut to a uniform length of 20 cm. The boxes filled with bedding materials, tightly packed in parallel rows. Materials can be bundled to facilitate the filling of the boxes. The packed materials loosen when soaked in water, soaking can be done for 3 hrs. After soaking straw becomes brown in colour and banana leaves becomes transparent. After this boxes are removed so that water can drain out. Then spawn is placed at the edge of soaked substrate. Then placed back in boxes in a incubation room at the temp. of 35° to 38°C, RH at 85 per cent. Boxes are covered with plastic sheets and placed under shade outdoors. With in 5-7 days spawn runs on the substrate and then pinheads begin to form along the open edges of the box.

2. Non Pasteurized Outdoor Beds by Bundles

The bundled paddy straw is soaked in water overnight and piled into heaps of three to four layers, thickness of the pile can be upto 30-60 cm thick. Thickness of the bed depends upon the weather. During winter thickness can be up to 60 cm so that heat will be produced and spawn can run easily, during summer season thickness can not be more than 50 cm because heat will be generated by the high temperature and spawn will not run. Each bed is covered with plastic sheets to maintain humidity. To get high yield of mushroom, quality of spawn, good substrate and cultural management conditions are required.

Indoor Beds (Pasteurized)

In this method, substrate is pasteurized directly inside the mushroom house.

Material used in cultivation is:

Cotton waste or chopped paddy straw, wooden frames of about 92 x 28 cm (for compost top and bottomless boxes) water $CaCO_3$ 2 per cent W/W of substrate rice bran and chicken manure (5 per cent W/W) or sewadge sludge (5 per cent to 10 per cent W/W) or NPK chemical fertilizer (1 per cent W/W). For preparation of compost a wooden frame is placed on the ground and paddy straw or cotton waste straw raw materials are packed inside. The materials are trecked as they are added, to help mixing and absorption of water. When frame is completely filled, it is pulled up to a height of 15-20 cm and another layer of compost is added in top the first layer. A finished pile of compost usually consists of four to six layers and is about 70-90 cm in height.

Pile is covered with plastic sheet and fermentation takes place with in two days. Then 2 per cent rice bran is mixed in it. Water is added if necessary. The mixed compost is remixed and covered with plastic sheet. Fermentation continues for another two days. Temperature of the room to be maintained 60°C for 2 hrs. With the help of steam supply of fresh air, it is lowered to 52°C for 8 hours. Then temperature is maintained at 34°C-36°C by closing the steam valves. Depending upon the outside temperature, it takes 12-16 hrs to lower down. At this stage beds are ready for spawning.

Spawning

For spawning, bed temperature about 36-38°C, air temperature of 35°C is needed.

The amount of spawn to be inoculated is 0.4 per cent (wet weight) and 1.4 per cent (dry weight) of the compost. Pure cultures are taken out from bottles and placed in the compost in small pieces (size up to peanuts) at a distance of 12 to 15 cm apart. From that, spawn is covered with the compost and covered with plastic sheet. Room temperature of the room is maintained at 32 to 34°C for the running of mycelium on the compost. After 3-4 days full growth of mycelium will takes place under proper conditions of temperature and by good quality of compost. Little amount of water supply and fresh air is required for running of mycelium. After three days more light and air is required. After 5 days small primordial can be seen on the compost. First flesh is ready to harvest with in 2 days of primordial formation(Figure 7.7).

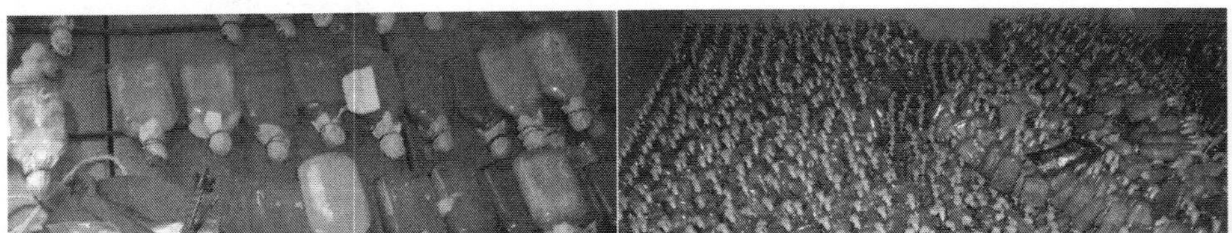

Figure 7.7: Spawn preparation

Harvesting

The first crop of mushroom can be harvested within 10 days of planting of spawn. 85-90 per cent yield of crop can be harvested during first flush. For the next 3-5 days watering of the compost is required. Stable conditions to be maintained in the growing room during harvesting. Spraying with superfine mist will maintain the desired humidity in the room and protect the substrate from drying. Ventilators opening and closing can be help in maintaining the temperature of room. After 2-3 days second flush can be collected but yield will be 70-75 per cent. *Volvariella* is best harvested at egg stage when the volva or universal veil has not broken. Rajapakse (2011) reported that indoor cultivation with cotton waste compost substrate gives the highest yield and paddy straw compost outdoor cultivation. Temperature and relative humidity can be maintained easily in room (Figure 7.8-7.9).

Outdoor cultivation does not provide the necessary environmental controls as indoor cultivation method.

Cost Effective

One bed of paddy straw mushroom prepared from paddy straw costs around Rs.20-22 Rs and it can yield 1-1.250 kg fresh mushroom, which cost about Rs.50-60 Rs net saving of Rs.30-38/bed. 9 ton of paddy straw in one month can generate an income of Rs.20,000/month.

Figure 7.8: Croping & Harvesting

Figure 7.9: Post harvest management of *Volvariella*

Guidelines for Good Quality Crop

- Moisture of compost must be maintained around 60 to 65 per cent
- During spawning, the temperature of compost to be 35°C
- After spawning, Ventilation is not required
- Plastic sheet to be removed after 5-6 days and regular sprinkling of water is required.

Trouble Indicators

- It bed is compact and inadequately beaten mycelium will not be able to run.
- Spawn should not be contaminated, otherwise contaminant will kill the spawn.
- Due to scarcity of water drying of mycelium takes place.
- Excessive use of ammonia cause foul smell.
- Degenerated spawn and too old spawn causes failure of fruiting body.
- Insufficient O_2 and excessive CO_2, poor Ventilation can be a cause of death of young mushrooms.
- Due to excessive use of nitrogen, growth of *Coprinus* can take place.

Value Addition

Paddy straw mushrooms is very common in North Eastern region, West Bengal and Orissa. Women used to go for hunting in woods. But one should be careful while plucking it due to unawareness, poisoning can takes place. People are generally consuming it in vegetable form with rice.

Disease of *Volvariella*

This mushroom being very fast in growth usually does not infected by diseases if it is handled properly. *Coprinus* sp., *Trichoderma* are common competitor moulds, which appear due to inefficient substrate pasteurization treatment or excessive use of nitrogen sources or watering. Among the insect pest aphids and mites are the common insect pests of this mushroom. Nematodes infestation also causes crop losses in certain cases.

Answer the Following

Q.1. Discuss the classification of *Volvariella* Mushroom.

Q.2. Discuss about the various methods of cultivation of *Volvariella* Mushroom.

Q.3. How Indoor cultivation is better than outdoor cultivation ?

Q.4. Mention various species of *Volvariella*.

Q.5. Explain with suitable diagram the life cycle of *Volvariella* Mushroom.

Q.6. What can be the cost benefit ratio by growing *Volvariella* Mushroom?

Q.7. Which disease can be seen on substrate of *Volvariella* and how it can be controlled.

Q.8. How bed can be prepared for growing of *Volvariella* Mushroom ?

Chapter 8

Cultivation of Black Ear Mushroom (*Auricularia* sp.)

Auricularia auricula – Judae is known by various common names *i.e.* Jelly ear or wood ear. It is a edible Auriculariales fungus found world wide. The fruiting body is distinguished by its ear – like shape and brown colouration.

Scientific Classification

Kingdom	–	Eumycota
Divison	–	Basidiomycota
Class	–	Basidiomycetes
Order	–	Auriculariales
Family	–	Auriculariaceae
Gemes	–	*Auricularia*
Species	–	*auricula – Judae*

The fungus can be found throughout the year in temperate regions world wide. It can grow upon the dead and living wood. It have medicinal properties can be used for the cure of sore throats, sore eyes and jaundice as an astringent. It is very popular in China and very commonly grown there. Australia exports large amount of *Auricularia* from China. This fungus is a popular ingredient in many

China dishes, such as hot and sour soup and also used in chinese medicine. In Ghana, people are using it as a blood tonic. According to modern research it has antitumor, hypoglycemic, lower cholesterol properties and anticoagulant in nature.

Figure: 8.1: Fruiting body *Auricularia*

Know Yourself

1. **What is the common name of *Auricularia auricula ?***
2. **In which country it was grown first time?**
3. **Why it is given the name *Auricularia auricula judae?***
4. **Mention the classification of *Auricularia.***

Taxonomy

In the scientific literature, species was first time mentioned as *Tremella auricularia* by Carl Linnaeus in 1753 (species plantarum). Later in 1789 Jean Baptist described it as *Tremella auricularia Judae.* The genus *Tremella* is reserved as for fungal species that lie as parasite on other fungi The species was given the name *Auricularia auricula Judae* in 1988 by Joseph Schrotar (Schrotar, 1988). *Auricularia* is a Latin word meaning ear and Judae meaning of Judas. This name was criticised by the mycologist Curtis Gato Lloyd – According to him *Auricularia auricua Judae* is a cumbersome. Despite all controversies *Auricularia auricula Judae* is recognised name for the species by many sources (Schrotar, 1988). Mycologist

Mary F. Barrottee attributes "such multiplication of names" to the wide distribution of the Judaes "ear, its ability to grow upon many different kinds of decaying wood and to its great variation in size, colour and shape. This fungus is associated with Judag Scariot because of the belief that he hanged himself on an elder tree after his betrayl of Jesus Christ. Folklore suggests that the ears are Juda's returned spirit (Geoffrey 2003).

Fruiting Body

The first body of *Auricularia* is normally 3-8 centimeters, but can be 12 cm in size. It is very different in shape, typically like a floppy ear, fruit body can also be cup shaped. It is attached to the substrate laterally by a short stalk. During fresh conditions, fungus body is tough, gelatinous and elastic in texture, after drying it becomes hard and brittle. The outer surface is a bright reddish brown with a purplish tint, often covered by tiny, downy hairs of a grey colour. It can be smooth, as is typical of younger specimens, of undulating with folds and wrinkles. The colour becomes darker with age. The inner surface is a lighter grey brown in colour and smooth, sometimes it is wrinkled, making it appear even more ear like. The spores of *Auricularia auricula* are long and sausage shaped, ranging in size from 16-18 micrometer long, 6-8 um thick. The spores are white, cream, yellowish and hyaline (Figure 8.2).

Figure: 8.2: Fruiting body of *Auricularia auricula*

Life Cycle of *Auricularia* sp.

Basidiospores are haploid and grow into a haploid mycelia under suitable environmental conditions. Hyphae from two haploid mycelia of opposite mating type undergo plasmogamy, changes into a dikaryotic mycelium that grows faster. It is followed by karyogamy occurs in the terminal dikaryotic cells. Each cell swells to form a diploid mycelium. In turn Basidium undergoes meiosis and yield four haploid nuclei. The basidium then grows four sterigmata and one haploid nucleus enters into each appendage and develops into a basidiospore. Then mature basidiospores are ejected from the sterigmata by a process of ballistospory. Then the spores drop below the cap they are dispersed by the wind (Figure 8.3).

Figure: 8.3 *Auricularia* fruiting bodies growing on the substrate

Answer the Following

1. **Discuss the shape and texture of *Auricularia* mushroom.**
2. **How the structure and texture of *Auricularia* is different from other mushrooms?**
3. **Define Ballistospory.**

Auricularia polytricha is an edible mushroom. The mushroom is cultivated in tropical regions because its mycelium can grow at temperature ranging from 10 to 40°C (Jonatha *et al.,* 2009). Unlike other mushroom species, cultivation of *A. polytricha,* is easy and fast to yield fruiting bodies. It does not require expensive facilities. Various forestry and agricultural waste have been used as substrates for cultivation of this mushroom. *Auricularia polytricha* has been extensively cultivated in Indonesia. Using substrates made from various type of lignocellulosic material.

Indonesia is one of the tropical countries which have many forestry sites and highly active wood industries,. Using wood waste as cultivation substrates for *A. polytricha,* is an effective utilizing method. Mycelial growth and fruiting body formation are greatly affected by tree species and quality. Irawati *et al.,* (2012) investigated three tree species, as substrate for the cultivation of *Auricularia* wood meal of 3 tropical hardwood species *i.e. Falcataria moluccana, Shorea* sp. and *Tectona grandis* were used as basal cultivation substrates. The fastest mycelial growth was found in the substrate made of *Shorea* sp. and the highest glucosamine content was found in the substrates made of *Shorea* sp. and *Falcataria* sp.

Do you Know

1. **Discuss about the substrates on which *Auricularia* mushroom can be grown.**
2. **Discuss the life cycle of *Auricularia* mushroom.**

Growing Region of *Auricularia*

Auricularia sp. is widely cultivated in China, Taiwan, Thailand, Philippines, Indonesia and Malaysia. Black fungus or wood ear mushroom are widely considered to be the earliest cultivated mushrooms (Tang *et al.,* 2010). Production of wood ear accounts for about 60 per cent of the world's total output of mushrooms. Annual production of *Auricularia* sp. in China alone reached nearly 3.6 million ton in 2010 making them the second most widely cultivated mushroom in that country.

Auricularia polytricha and *A. auricula* are two main species which are grown/ cultivated commercially. This mushroom has special quality of retaining its characteristic- crispness on cooking.

Flow Chart for Cultivation of *Auricularia*

Cultivation

Take good quality of wheat/rice straw

↓

Soak the straw over night for 16-18 hrs then drain out the water.

↓

Mix 4-5 per cent wheat or rice bran on wet weight basis

↓

Fill 2 kg PP bags and autoclave at 15 lb pressure for 2 hrs.

↓

On cooling, spawn is added into substrate @ 2 per cent aseptically and incubate at 25-30°C for spawn run for 20-25 days.

↓

↓

Give slits in the bag and hang the bags for fruiting at 25°C and relative humidity 85-90 per cent.

↓

Daily aeration of 2 hrs and spraying of water.

↓

After 5-7 days fruiting bodies will appear in the form of Pin heads and with in 4-5 days these will be ready for harvest.

↓

Fruiting bodies can be marketed fresh as well as in dried form.

Cultivation involves two important steps like spawn preparation and fruiting body formation.

Fruiting body formation

↓

Log method cultivation | Artificial log method | Polybag method

Artificial log method ↓

Cultivation in Sawdust | Cultivation in compost

a) Log Method of Cultivation

Small logs can be made from tree trunks of 30x15 cm size. The cut surfaces can be smeared with Bordeaux mixture to prevent the attack of wood-decaying fungi. The logs can be inoculated by spawn of *Auricularia* after boring holes of 1 cm deep with a hand drill. The inoculated logs are covered with bark and smeared with wax in order to avoid decay. The cultured logs can be pilled up, as these takes longer time to produce fruiting bodies.

b) Artificial Log Method

i) Cultivation in Saw Dust

Material Required:

Saw dust – 78 per cent

Rice bran – 28 per cent

Calciumcarbonate – 1 per cent

White sugar – 1 per cent

Saw dust and bran has to be soaked in water for a day. Excess water to be drained out from the husk and bran, spread it and make it dry in shade.

Above mentioned ingredients after mixing are filled in PP bags. The mixture has to be sterilized in an autoclave. After autoclaving, these bags have to be inoculated with spawn. Inoculated bags are then incubated at 20-25°C for spawn run. After 15 days bags are cut open for inducing fruiting bodies (Stametes, 1993).

ii) Cultivation in Compost

For compost making, saw dust – 78 per cent, Rice bran- 28 per cent, Calcium carbonate – 1 per cent and white sugar – 1 per cent are required. Compost formation usually takes 6-7 days. All ingredients are mixed thoroughly and moisture level is to be maintained around 65-70 per cent. Pile the material in a Pyramid shape; turn upside down all the material at the interval of 2 days. After the compost formation, it can be sterilized by 0.1 per cent formalin spray, cover it and left it for 2-3 days. The inoculation of the spawn will be done in the compost by thorough method. Then it will be filled in bags and will be kept for incubation and fructification at particular temperature (Onimio, 1976).

c) Poly Bag Method

Take good quality of wheat straw, soak it in water for 16-18 hrs. After that drain out excess of water and add 5 per cent wheat bran (w/w). Fill 1 to 2 kg of substrate in polypropylene bags depending upon their size and autoclave at 15 ± lb pressure for 1.5 hr. After cooling of substrate, spawn is to be added at the rate of 2 per cent and incubate the bags at 25-26°C for mycelium to colonize, humidity to be maintained at 85-90 per cent by spraying water in the room. Give 1-2 hrs diffused light and aeration also. After 10-12 days fruiting bodies will be emerged out and harvesting can be done after 4-5 days. One can harvest 1-1.5 kg fresh mushroom from one kg dry straw in 3-4 flushes. Mixing of rice bran and wheat bran at the ratio of 1:3 was found to induce faster growth of *A. polytricha* beds (Devi *et al.,* 2013).

Medicinal Importance

Auricularia auricula Judae and *A. polytricha* are very popular in China as medicine. A soup containing the *Auricularia* sp., is used medicinally for dealing with cold and fevers by reducing the heat of body.

Two Glucans isolated from the *A. auricula* showed potent anti-tumour properties. Polysaccharide extracted from *Auricularia judae* has hypoglycemic effect. Polysaccharide reduces plasma glucose, insulin, urinary glucose and food intake (Yhan *et al.,* 1998).

It is very popular in Vietnam. The annual production of *Auricularia* species world wide is the fourth highest among all industrially cultivated culinary and medicinal mushrooms, China estimated output was roughly 1.655 million tonnes, *A. polytricha* replaced *A. auricula Judae* in international trade (Tang *et al.,* 2010).

Uses

Auricularia auricula Judae has soft, jelly like texture. Initially it was not regarded as an edible fungus. Till 19th century in Britain it was not in the list of edible fungi. It has a mild flavour can be used for mixed mushroom recipes, still it is considered bland in the taste (Conta *et al.,* 2008). It can be dried and rehydrated and it swells to very large size. *Auricularia* should not be eaten raw, after thoroughly cooking it can be consumed.

Nutrient content in the *Auricularia*: 10 gm of dried mushroom contains 370 kcal, 106 gm of protein, 0.2 gm of fat, 65 gm of carbohydrate, 5.8 gm of ash and 0.03 per cent mg of carotene. Fresh mushroom contains 90 per cent moisture. Dried specimens may be ground up into a powder and used to absorb excess liquid in soups and stews, due to its rehydration nature into tiny fragments (Tomblin, 2007).

Another chemical extracted from the species was an acidic polysaccharide made up of mannose, glucose, glucuronic acid and xylose which showed anti-coagulant properties. It is used to lower cholesterol, levels generally and in particular is one of fungi that reduce the level of bad cholesterol (Francia *et al.,* 1999).

Answer the Following

1. Explain various methods of cultivation of *Auricularia* mushroom.
2. How *Auricularia* mushroom is important medically?
3. Mention the nutritive value of *Auricularia* mushroom.
4. Discuss the history of naming of *Auricularia* mushroom.

Chapter 9

Diseases of Mushroom

The population of the world is increasing at an alarming rate (1.8 per cent annually), it was estimated to reach 7.8 billion people as of March 2020 (World Population Clock, 2019). It took over 2,00,000 years of human history for the world's population to reach 1 billion (World Population, 2018) and only 200 years more to reach 7 million.

About 3 billion human beings are living today in absolute poverty. Their number is more in under-developed than developing countries due to steep rise in population. There is an acute malnutrition as a consequence of lesser protein intake. Protein is essential for good health, the recommended dietary allowance (RDA) for protein is a modest 0.8 grams of protein per kilogram of body weight. FAO has recommended edible mushrooms as an alternative source of protein nutrition. Recent technological changes in the production of cereal crops have resulted in availability of large quantities of crop residues.

(Adaptation of the advanced pest management system to control damage caused due to pest infection in mushroom, can further result in higher mushroom yield, which will ultimately aid the market growth of mushroom).

Like any other living organisms, mushrooms are subject to attack by diseases. These can be seen on specially prepared compost which favour the selective growth of the mushroom mycelium. If composting is not carried properly, moulds develop, some of which can reduce the yield, that depends upon their incidence and stage of infection. Presence of moulds in compost indicates that composting has not been done properly, these are referred as indicator moulds, weed moulds compete with mushroom for compost nutrient and known as competitor moulds. Most of the pest and disease problems can be avoided by pasteurized composed under hygienic conditions, there by emphasizing the need for use of properly pasteurized compost and casing soil. Main abiotic factors are temperature, RH, carbon dioxide, excess moisture and presence of toxic chemical in atmosphere and substrate.

Know Yourself

1. **Define moulds?**
2. **How many types of disease can be in or on mushrooms?**
3. **Define competitor moulds.**

Mushroom growers, especially the seasonal growers face multiple problems of fungal, bacteria, insect pest and nematode infection. General distribution of various pathogenic fungi and moulds are as follows:

a) Occurring Mostly on Compost

Chaetomium globosum and C. olivaceum (olive green mold), *Coprinus* (Ink Caps), *Aspergillus, Penicillium* and *Trichoderma* (green mould). *Mucor* and *Rhizopus* (Black moulds) *Myricoccum, Sporotrichum, Sepedonium, Dedocephalum* sp., *Fusarium solani, Cephalosporium roseum, Gliocladium* sp. and *Papulaspora byssina etc.*

b) Fungi Occurring on Compost and Casing Soil

Scopulariopsis fimicola (White plaster mould), *Papulospora byssina* (Brown plaster mould) *Geotrichum candidum, Sporondonema purpuresceus* (Lip stick), *Diehliomyces microsporus* (False truffle) and green moulds.

c) Fungi Occurring on and in Casing Soil and on Growing Fruiting Bodies

Peziza ostracoderma (Cinnamon mold), *Mycogyne perniciosa* (Wet bubble), *Verticillium fungicola* (dry bubble), *Dactylium dendroides* cobweb, *Fusarium* and Green moulds.

d) Fungi Attacking on the Fruit Bodies Only

Fusarium sp., *Verticillium* and *Cladobotryum* sp.

e) Bacterial and Viral Infections

Answer the Following

1. **Discuss on which substrates moulds can be found.**
2. **Mention some common organism found on compost.**
3. **How moulds can be controlled?**

Common Pest Disease

1. Fungal Causal Organism (*Dactylium cladobotryum*)

Symptoms

Webbed, cottony, mycelial growth on the surface of casing and on mushrooms; mycelium is usually white but can be grey, pink in colour infected mushrooms will be soft and watery in texture.

Growth of the mold is restricted to the casing layer, disease can spread very fast, if left untreated (Figure 9.1).

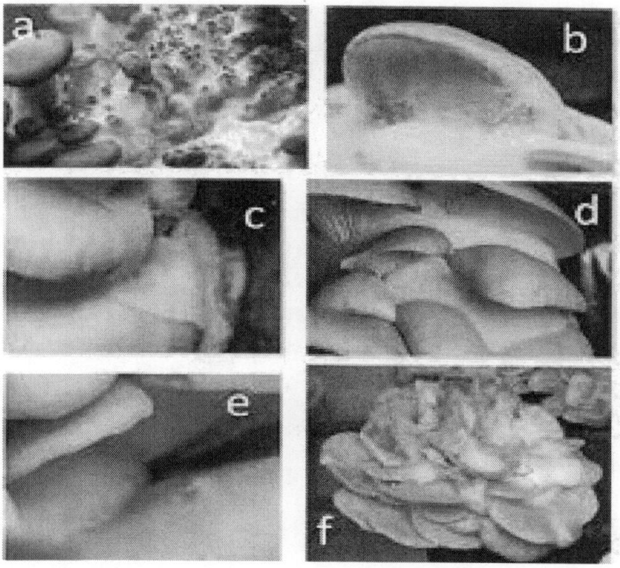

Figure 9.1: *Dactylium cladobotryum* on *Pleurotus* fruiting body.

Management

Control of the disease can be done through proper sanitation and cultural control methods. Casing should be kept clean and sanitized; all tools and equipments should be sanitized, beds must be free from mushroom debris, humidity in the room must be maintained properly.

2. Green Mold (*Trichoderma, Penicillium* and *Aspergillus* sp.)

Causes blotch or mildew.

Symptoms

Green colonies with whitish border appear on casing soil and on compost. It also attacks grain in spawn bottles. Greenish lesions, spots and discolouration appears on the pileus parts and later on it appears on stipe parts also. Due to infection spawn running is inhibited and fruit formation also did not takes place.

Trichoderma konigii grows as cottony growth of mycelium over the casing. It also covers the furit bodies with a cottony web producing wet rot. Purpule brown spots formation takes place on the fruiting body, with a dry cracks on the surface. Loose green colour of spore develop on affected tissues. *Trichoderma viride* is associated with decaying organic matter in the casing soil. It produces large dry brown, sunken lesions on cap and stem. *T. hamatum* spores are dark green, produces less spores which are pale green and smooth. The infection generally comes through compost and from air (Figure 9.2).

Figure 9.2: Green Mould (*Trichoderma*)

Management

Disease can be controlled by the spray of benomyl or bavistin (25-30 ppm) sprays.

3. Dry Bubble

Causal Organisms: *Verticillium fungicola*

It is also known as brown spot, fungus rot or *verticillium* disease. It is a wide spread, serious and destructive disease occurs in almost all the mushroom growing countries including India (Seth *et al.,* 1973). Whitish mycelia appears in rainy season, which later on changes into greyish yellow in colour (Figure 9.3).

Later on due to infection fruiting bodies becomes deformed, shabby deformed stipes with tilted caps. In fully developed sporophores, brown spots can be seen but later on coalesce to form large blotches. Fungus is soil borne, spores can survive in soil for a year. Primary infection comes through contaminated casing soil, by workers, flies and mites from spent compost. Secondary spread of infection takes place through water splashes by flies. High humidity, lack of proper air circulation resulting in condensation of moisture on fruit body surface. All the commercial strains of *A.bisporus* are susceptible.

Figure 9.3: Dry Bubble (*Verticillium fungicola*)

Management

Use of sterilized casing soil, proper disposal of spent compost and better hygiene is essential to avoid primary infection (Sohi, 1986). Spraying with Dithane from 0.2-0.5 per cent, 1± lb in 100 gallons of water once before first break (Yoder *et al.,* 1950). Benzimidazole fungicides was very effective initially, but the fungus developed resistance (Wuest *et al.,* 1974). Compost must be pasteurized properly, regulation of RH and temperature during cropping with good hygiene and proper disposal of spent compost can help in control of disease.

4. Wet Bubble

Causual Orgaism: *Mycogone perniciosa*

This disease is also known as bubble, white mould or *Mycogone* disease. This disease is reported from England, USA, France and India. It adversely affects yield and quality of mushroom, white, fluffy mycelium patches can be seen on the surface of casing soil infected mushrooms have swollen stalks, smaller caps and surface is covered with white mycelium. When infection starts at the early stage of casing, mushrooms become distorted, necrotic and brown in colour. Amber to dark brown drops of liquid oozes or accumulates on the surface of undifferentiated diseased tissues. The contaminated casing soil results in early infection learning to serious yield loss or even crop failure. The spores are carried by flies, tools and water splashes (Figure 9.4).

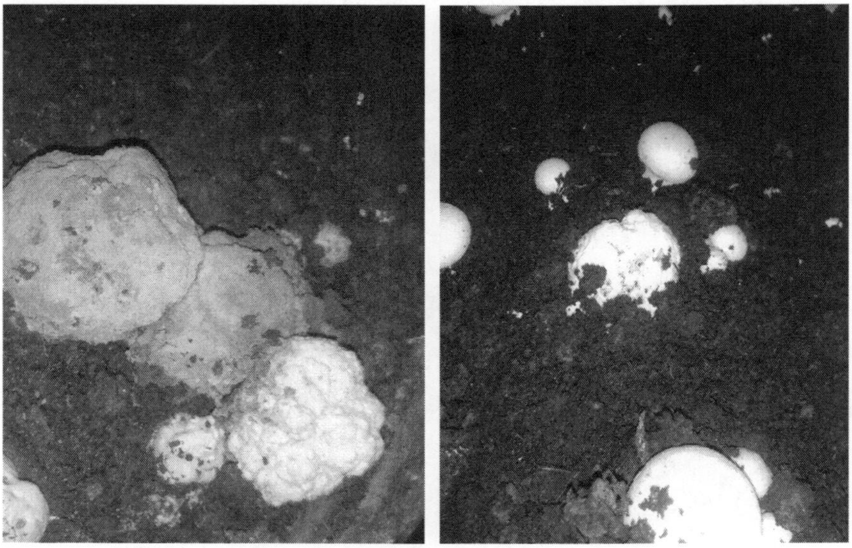

Figure 9.4: Wet Bubble (*Mycogone*)

Management

Casing soil must be sterilized properly. Good hygiene must be maintained. Diseased mushrooms must be picked or destroyed before watering. Fumigation of casing soil with 1-2 per cent formaldehyde can be done. Spray of Benomyl with 70z in 20-50 gallans can help in controlling the diseases (Atkins and Atkins, 1971).

5. Mildew Disease

Casual Organism: *Dactylium dendroides* – the conidal stage of *Hypomyces rosellus*).

It is also known as cobweb disease or soft decay disease. Its cosmopolitan in distribution from India it was reported by (Seth and Dar, 1989). Generally it attacks later flushes but it can be serious during its favourable conditions of spread. Irregular, white patches with loose mycelium can be seen on casing surface. Eventually, it turns brown or rosy red. Fruit bodies can be attacked by the fungi at any stage, mycelium varies from white to brown in colour. Casing soil is the primary source of infection. Sexual stage has been frequently observed on decaying dried bodies of mushrooms. Drop in temperature and excess of moisture over casing surface favour disease.

Management

Sanitisation of cropping rooms and trays by 1.5 per cent formaldehyde minimize chances of infection. Benizimidazole sprays is effective, dusting with Zineb and spray of 20 per cent PCNB or 75 per cent Calcium hypochlorite is also recommended (Stoller *et al.,* 1956). Excessive watering should be avoided.

Figure 9.5: Mildew disease (*Dactylium dendroides*)

6. Brown Plaster Disease

Caused by – *Papulospora byssina*

This disease has been reported from USA, UK, Germany, China, Brazil, Australia and India (Munjal *et al.,* 1974). Infection results in yield losses and delayed cropping. It causes 90-92 per cent reduction in yield of white button, complete crop failure of oyster mushroom has been reported from Kasauli. Infection is more common at almost all the farms using unpasteurized compost. Incidence is more when cultivation is done in polythene bags, filled with large amount of compost. Initial infection is in the form of white mycelial growth on the exposed part of compost. It is found in the compost due to moisture condensation. It develops into large dense patches changing colour shades from light brown to cinnamon brown. Mycelium becomes brownish, septate and small clustures of brown colour can be seen. These are interwoven with a net work of hyphae and set free with death of mycelium (Figure 9.6).

Figure 9.6: Brown Plaster (*Papulospora byssina*)

Management

Fungus can be killed at 140°F. Proper pasteurization of compost and casing soil, can be helpful in control of disease. Use of wet compost can be avoid.

Infected patches can be treated with 0.2 per cent formaldehyde spray.

Benomyl and Bavistin (0.05 per cent) is effective against mycelial infection.

7. White Plaster Mould

Casual Organism: *Scopulariopsis fimicola*

Infection is quite similar to brown plaster but remains white throughout its life. It was reported from India by Bhardwaj *et al.* (1989). Mycelium is white, septate, fasciculate, conidiophores short, mostly branched and produced at apex in cluster of sporogenous cells. Infection is favoured by wet and wrongly prepared compost and high temperature during spawn run (Figure 9.7).

Management

Spray of Dithane-Z-78 or Bavistin (0.05 per cent) or formalin (2 per cent) are effective.

Figure 9.7: White Plaster *(Scopulariopsis copulariopsis fimicola)*

8. False Truffle

It is known as a parasite or competitor mould. The infection was first noticed in USA, UK and then in India (Sohi and Kumar, 1965).

Symptoms

Symptoms of this disease can be seen in the month during April-May in hilly areas, but in plains it can be from February-March. Lack of proper temperature control can be the reason for this disease (Figure 9.8).

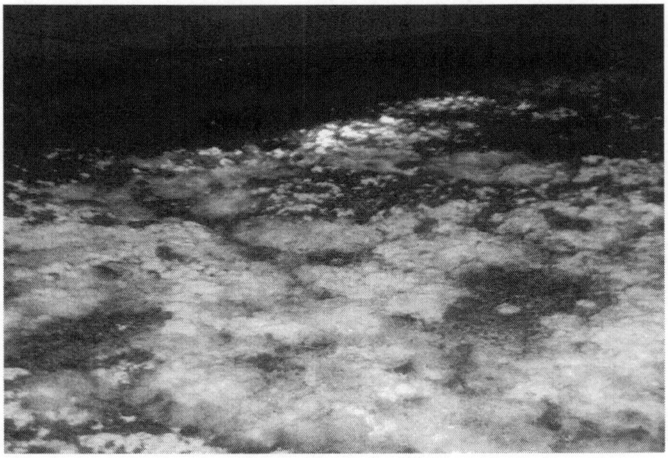

Figure 9.8: False Truffle

Favourable temperature of its growth is 22-28°C. Infection can be seen as a small wefts of mycelium varies in colour from white to green in compost. Gradually mycelium changes into small, hard bodies like a to small brain of peeled walnuts. Due to infection spawn becomes soggy and yield of mushroom is affected. In the ascocarp, numerous stalked asci formation takes place, which is reddish brown, oval shaped, spherical and short. Ascospores are spherical, sulphur coloured, with distinct oil drop. These are the source of primary infection and can survive for long period in the soil and spent compost ascospores can survive up to a temperature of 82°C.

Management

Maintenance of temp. of 24°C during spawn run and below 18°C during cropping, with proper aeration is essential to avoid infection strict hygienic conditions to be maintained. Infected trays and bags must be removed or destroyed before its spread. Infected trays can be treated with Sodium pentachlorophenate. Use of copper sulphate can be helpful in controlling the disease. Stoller (1943) reported that fumigation with ammonia at a conc. of 0.15 gm per litre can helpful in preventing the infection for next generation.

Bacterial Disease

It is also known as brown blotch or bacterial spot, caused by *Pseudomonas fluorescens.*

Symptoms

It is a very serious disease and cosmopolitan in distribution. Yellow to brown spots can be seen on the cap, which turns into dark brown in colour, if infection is severe. Sunken pits filled with slimy contents can be seen. On the stipes brown stripes formation takes place, pin heads cracked radially. Infected tissues changes into water soaked area and grayish in colour; cap is greasy in touch. High relative humidity of more than 85 per cent, temperature beyond 20°C and fluctuation in temperature during cropping system results in condensation of moisture on fruit body surface, favours the development of infection. Infection affects the pin head formation so yield and quality of mushroom is affected. Bacterium is omnipresent in distribution so source of primary infection can be compost, soil/or containers. It can spread by flies, nematodes, pickers or by water splash.

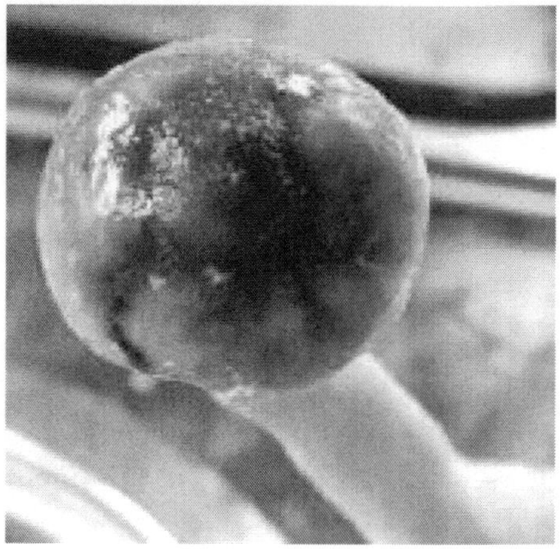

Figure 9.9: Bacterial Blotch (*Pseudomonas fluorescens*)

Management

Proper air circulation after watering is required to avoid infection.

Early and careful trashing is required. Casing should be done properly, taking care of sanitation and sterilization will help in control of disease.

Spraying with chlorinated water 500 ppm from the beginning of first flush, twice a week can help in disease control. Bacterial antagonist pathogen can help in control of bacterial disease (Figure 9.9).

Viral Disease

These disease are named as Die-back, watery spike, La-France and X-disease. Those are the most serious and fast multiplying disease of cultivated mushrooms. These diseases are reported from USA, England, Netherlands, Australia and India (Tiwari and Singh, 1984). It is caused by Mushroom virus X (MUX) which is dsRNAs.

Symptoms of the disease are available and not easy to recognize. Growth of mycelium is very sparse, suppression of development, rapid necrosis, dwarfing and early maturity of sporophores takes place. In the fruiting body long stalk and short cap formation takes place, premature opening of veil, bending, splitting and swelling of stalk can be noticed. During severe infection absence of mycelium formation can be noticed. Failure of crop or loss in yield can be seen. Infection mostly comes from the diseased compost, absence of mycelium formation can be noticed. Failure of crop or loss in yield can be seen. Infection mostly comes from the diseased cultures and spores from infected mushrooms. Quick germination can be seen of diseased spores. Mycelium can survive in wood or trays and can be transmitted to the healthy compost. Browning of fruiting body also takes place at the later stage.

Management

Strict hygienic conditions must be followed. Disinfect all trays, growing rooms, implements and machinery after operation including work areas and growing rooms. Burning of spent compost can be helpful in controlling the disease. Proper filters can be arranged in room to avoid entry of virus.

Weed Moulds

Brown Mould (*Peziza ostracoderma*)

Various cup shaped, brown to dark brown fruit bodies can be seen on compost surface. High humidity and low ventilation in growing room supports its growth. Brown mould is caused by *Oedocephalum* sp. It is due to the ammonia in compost.

Ink Caps

Ink caps (*Coprinus atramantarius, C. logopus, C. comats, C. fimetarius*). These appear during spawn run and cropping and show profuse unrestricted white mycelial growth followed by appearance of fruit bodies in large numbers. It inhibits spawn run and fruiting, it becomes serious with wet compost, young fruit bodies of the mould can be picked to avoid its further spread (Figure 9.10).

Figure 9.10: Ink Caps (*Coprinus atramantarius*)

Lip-Stick Mould (*Geotrichum candidum*)

It produces red colonies with white border in compost or casing and inhibits spawn run. Air is the source of infection.

Mat Disease: (*Chrososporium luteum*). Small white colonies becoming yellow appear underneath casing.

Yellow mould: (*Sepedonium* sp.). It produces small yellow colonies in compost and affects production.

Most of these weed moulds can be controlled through good hygiene, proper compost preparation followed by pasteurization and providing desired growing conditions. Spent compost needs to be disposed carefully away from the cropping site (Sohi, 1998). Ecologically manipulation of environmental factors such as temperature, humidity and ventilation. Biological use of antagonistic organisms through in corporation of biocontrol agents and organic amendments. Spray of chemicals can also be helpful. As chemicals are not eco-friendly so it can be replaced by biological methods.

Nematodes are another class of pest mushroom which can be parasitic as well as saprophytic. Parasitic nematodes are dangerous as they cause failure of mushroom crop. Saprophytic nematodes eats upon the vegetable matter of compost and casing soil, loss is negligible. *Ditylenchus myceliophagus* and *Aphelenchoides composticola* exclusively thrive on mushroom mycelium and destroy the spawn completely.

Answer the Following

1. **Explain the disease on mushroom and its compost caused by various fungi.**
2. **Discuss about the bacterial diseases on the mushroom, compost and on casing.**
3. **Mention various diseases caused by insects during mushroom cultivation.**

Chapter 10

Current Status of Mushroom Production in India

India's agriculture as exist today is due to its long way journey that changes its image of being non-commercial and traditional in its method of farming. Business and investments opportunities in agricultural sector have jumped manifold as Government is taking care of this area. Under G20 Government is supporting new skills and entrepreneurship policy has been started under that mushroom cultivation and production is one of the agenda. In present diet conscious era, mushrooms are gaining importance and considered as a future functional food due to its medicinal and nutritional properties. In recent years the consumer demand for mushrooms have markedly expanded. Mushrooms are considered as a good source of protein as they are highly digestible (Pavel, 2009). They are excellent source of Vitamin D, as it is not available in other food supplements (Pehrsson *et al.,* 2003). Mushrooms are fat free, cholesterol free, gluten free and low in calories and sodium. Minerals are also available in the fruiting bodies.

Mushrooms are consumed as a medicine in Asian countries, researchers and medicine companies are working on its medicinal aspects. Mushrooms are used in Ayurveda and folk medicine in India (Jitendra and Vaidya, 2000). In India large amount of agrowaste is produced from the agriculture, approximately 620 million tons (Singh and Sindhu 2014). Burning of that agrowaste causes the air pollution. To avoid that condition and by adapting new technologies one can improve the economic power of the country by growing mushrooms. In India, mushroom is used as a non-traditional cash crop and commonly, cultivated species are white button mushroom, oyster mushroom, shiitake, paddy straw mushroom. From the last 12-15 years there has been exponential growth in production and quality of mushrooms in the country. Output is touching 50,000 tons now from less than 1000 tons in 1985 (Verma, 2002). Cultivation of oyster and shiitake mushroom is becoming very popular these days (Karthik and Hamsalakshmi, 2017).

Ask Yourself

1. **How growing of mushrooms can be part of sustainable Agriculture?**
2. **Mention main varieties of cultivated mushrooms grown in India.**
3. **How problem of malnutrition can be solved by growing mushroom?**
4. **How business can be generated by the mushroom farming?**

Button mushrooms accounts for approximately 95 per cent of total production and exports. Button mushroom is cultivated in temperate regions of Himachal Pradesh, Jammu and Kashmir. However the oyster, milky and paddy straw mushrooms are cultivated in tropical and subtropical regions (Jagdeesh *et al.,* 2018). The global mushroom industry has grown rapidly in recent years in terms of market value, beneficial effects and demand. Mushroom cultivation not only helps recycle agrowastes, but also fills the nutritional gap among a large population of India.

Mushroom cultivation has become a business and export oriented. The major export destinations for Indian mushrooms are Canada, US, Israel and Mexico. Mushrooms can be exported in two forms, fresh mushrooms and preserved mushrooms. It can be processed also, its demand is increasing over the past years. They can be canned, dried, packed in frozen forms, including its usage in food industry in mushroom pickle and sauces.

India's FDI (Foreign Direct Investment) policy aims to attract investment in technology to development and production of vegetables and mushrooms under controlled conditions (Annual Report 2017-2018). Efforts are being made by the Indian Government to improve Mushroom Research And Development and to encourage mushroom growers to develop advanced R&D technologies and policies. The R&D and biotechnology labs play a vital role in the optimization of the cultivation environment for the better yield. Such technological advancement and favourable government initiatives promise growth opportunities for the industry participants and the mushroom research (Jagdeesh *et al.,* 2018).

Brain Storming

1. **Mention some policies of Government in support of mushroom farming.**
2. **Discuss the role of FDI in promotion of mushroom cultivation.**
3. **India is maximum exporter of which mushroom.**

Status of Mushroom in India

Mushroom production and consumption are on the rise of the World. But in India response of growth is very slow. Mushroom industry in India is overwhelmingly focused on white button mushroom which is highly sophisticated and capital intensive activity. According to ICAR-DMR-Solan Data, button mushroom production in India is about 73 per cent followed by oyster mushroom which contributes about 16 per cent. There are two main types of mushroom growers in India, one growing white button mushroom round the year under controlled conditions and seasonal growers who are growing button mushrooms during the winter season in north-western part of India. The total white button mushroom produced in India from both seasonal and high tech cultivation units is estimated at 94676 metric tons. About 8500 metric tons of button mushroom is produced from seasonal growing units located in Haryana, Punjab and Maharashtra, are producing 43 per cent of total white button mushroom produced in country. Many medium to small scales units are local in the states of Gujrat, HP, TN, UK, UP and Goa (Tables 1 and 2).

Table 1: Some of the Medium to Large Scale Mushroom Production Units in India

Sl.No.	*Name of Mushroom Unit*	*Location*	*Production Capacity*
1.	M/s Himalaya Int. Pvt. Ltd.	Vadnagar, Gujarat	10000 TPA
2.	M/s Agro Dutch Mushroom Pvt. Ltd.	Chandigarh	8000 TPA
3.	M/s Himalaya Int. Pvt. Ltd.	Patna Saheb	4000 TPA
4.	M/s Zuari Agro Farm	Goa	4000 TPA
5.	M/s Triputi Balaji Agro Products Pvt. Ltd.	Baramati Pune	3000 TPA
6.	M/s Flex Foods Pvt. Ltd.	Dehradun	2600 TPA
7.	M/s Wiekfield Food Pvt. Ltd.	Haveli, Pune	2250 TPA
8.	M/s Welkins Overseas Pvt. Ltd.	Roorkee	500 TPA
9.	M/s Vikas Mushroom Farm	Solan, Shimla	500 TPA
10.	M/s Kulkarni Farm Fresh	Belgam, Karnataka	350 TPA

Source: Sharma *et al.* (2017).

Table 2: Some of the Small to Medium Scale Subtropical and Tropical Mushroom Production Units in India

Sl.No.	*Name of Mushroom Unit*	*Type of Mushroom*	*Location*	*Prodution Capacity*
1	M/s Prince Mushroom	Milky Mushroom	Erode Tamil Nadu	100 TPA

Sl.No.	Name of Mushroom Unit	Type of Mushroom	Location	Prodution Capacity
2.	M/s Chandera Sekran Mushroom Farm	Milky Mushroom	Villupuram TN	100 TPA
3.	M/s Siddharth Mushroom	Oyster Mushroom	Nashik	90 TPA
4.	M/s Mohan Prasad	Oyster Mushroom	Maharashtra's, Jamai Bilar	50 TPA
5.	Flora Agro	Oyster Mushroom	Thane Maharashtra	10 TPA
6.	Dhepe Mushrooms	Oyster Mushroom	Ahmednagar, Maharashtra	10 TPA
7.	Ruchi Mushroom	Paddy Straw Mushroom	Mayurbhanj Odisha	40 TPA
8.	Mrs. Jayanti Pradhan	Paddy Straw Mushroom	Kendupalli Odisha	10 TPA
9.	SM Mushroom Laramba	Paddy Straw Mushroom	Borgarh Odisha	10 TPA
10.	Mahadev Sahoo Barpali Bargarh	Paddy Straw Mushroom	Borgarh Odisha	10 TPA

Source: Sharma *et al.*, 2017.

There has been increase in production of mushrooms especially of oyster and paddy straw mushrooms in India during the last few years. The country's production in 2010 was 1.00 lakh metric tons, of which button mushroom account for 89.0 per cent of the total production, followed by oyster 6 per cent, milky 1 per cent and others 4 per cent, Punjab, Haryana, Uttar Pradesh, Uttrakhand, and Tamil Nadu are the leading producers of the mushroom in the country. The present production data, mushroom industry in India recorded an average, annual growth rate of 4.3 per cent. Well qualified Professional Mycologist, mushroom specialist, educated and unemployed population sector are contributing towards the cultivation of mushroom.

Mother compost unit and spawn labs in different agro-climatic area has been established all over India. The farmers who are growing vegetables now looking for another alternatives by cultivating the indoor mushrooms. Mushroom farming has become a very important Cottage industrial activity in the integrated rural development programme. The National Committee on the use of plastics in Agriculture (NCPA, 1982) has recommended green house technology for adoption in various regions in India. The indoor cultivation techniques and green house technology inspired villagers to set up mushroom units. The mushroom production in India in the year 2020 AD, would touch the magic figure of 6,00,000 tons per annum (Karthik and Hamsalakshmi, 2017). The cultivation of edible mushrooms not only helps in recycling of agro-wastes but also filling up the nutritional gap prevalent among large population of India. In addition, spent substrates can be used for biogas production, for fertilizer in agricultural fields and can also upgrade and uses as animal feed (Zadrazil *et al.*, 1992).

1. **Mention main centers of mushroom growing units.**
2. **Mention main labs growing spawns of various types of mushrooms.**
3. **Which states are leading producers of mushroom production in India?**

Globally, few of the mushrooms are brought under cultivation on a commercial scale. Experimentally, about 80 mushrooms have been grown but 20 are cultivated commercially, only 4-5 species are produced on the industrial scale throughout world (Chang, 1999). Though mushroom production in Asian countries started before 1000 years, cultivation of mushroom is a relatively new phenomenon in India (Sharma *et al.,* 2017). According to FAO the mushroom production has been increased over ten fold between the year 1969 to 2009. Inclusive more than 20 different types of mushrooms are cultivated on commercial scale. In India market is popularly dominated by *Agaricus bisporus, Pleurotus* sp. *Volvariella volvacea* and *Calocybe indica* (Dhar *et al.,* 2011). The most cultivated *Pleurotus* sp. are *P. sajor caju, P. florida, P. flabellatus* and *P. eous* and particularly *P. florida* are most popular (Directorate of Mushroom Research ICAR 2011).

In India button mushroom production share is 73 per cent and production systems are mixed type *i.e.* both seasonal farming as well as high-tech industry (Sharma *et al.,* 2017). The Directorate of Mushroom Research (ICAR-DMR) has developed 3 new strains of *A. bisporus* are NCS-100 (Single spore isolate),NCS-101, (SSI) NCH-102 (hybrid). Two improved strains of summer white *Agaricus bisporus* – NCB-6 and NCB-13 were introduced after evaluation trials and cultivation techniques standardization.

Oyster mushroom production share is 16 per cent and annual production 10,000 tons in India. It is popularly grown in the states of Orissa, Karnataka, Maharashtra, AP, MP, WB and in the North-Eastern states of Meghalaya, Tripura, Manipur, Mizoram and Solan (ICAR 2011). High yield strain of oyster mushroom PL-16-02 (66.07 kg/100 kg straw) was developed by the evolution trials.

Paddy straw mushroom *Volariella volvacea* high yield strains of Vv-16-02 (17.27 kg/100kg) Vv-16-03 and Vv-16-06 (17.70 kg/100kg) were tested in different locations in India.

Calocybe indica (Milky mushroom) is the first indigenous mushroom to be commercialized in the country. The maximum yield strains C1-16-03 (104 kg/100kg) and CI-16-01 (79.17 kg/100kg) on straw dry substrate were developed in Tamil Nadu and Uttarakhand (Annual Report 2016-17, ICAR-DMR).

Lentinula edodes also known as shiitake mushroom is an edible and culinary medicinal mushroom native to East Asia. In India cultivation of shiitake mushroom is negligible and very limited success due to the unstandardized cultivation techniques. National Research Centre for Mushroom (NRCM) Solan and Indian Institute of Horticultural Research (IIHR) Bengaluru are trying to develop cultivation technology based on locally available substrates (Sharma *et al.,* 2006). The high yield and maximum biological efficiency strain LE-16-04 (218 kg/100kg) of the dry substrate was developed at Haryana on saw dust based substrate (Annual Report 2016-17).

The black ear mushroom (*Auricularia* sp) is fourth most popular mushroom in the world. Unfortunately there are not any farms that cultivate this mushroom in India even though cultivation technology was standardized in 1986.

Reishi mushroom (*Ganoderma lucidum*) the commercial units are already in operation located in different regions of our country and producing the quality of mushrooms for export. *Ganoderma lucidum* is pharmacologically and economically potential mushroom the global trade of about two billion dollars; trade in India has crossed Rs.100 crores annually through imports from Malaysia and China (ICAR, Solan 2011).

Answer the Following

1. **Which improved strains of button mushrooms have been developed by ICAR-DMR?**
2. **Mention the improved varieties of Oyster, *Volvariella* mushrooms developed by ICAR-DMR.**
3. **What is the role of FAO in mushroom growing Industry?**

Spawn Production

With the setting up of the National Research Centre for Mushrooms at Solan, spawn was made available from the centre to small and marginal farmers.In the late eighties Sylvan from USA, started operations in India marketing spawn produced at their European centres. But they did not find the market to set up their own production unit. In India with R&D back up facility, the spawn companies are operating just multiply and sell it.

The availability of strain is limited to non hybrid strains of *A. bisporus* and no hybrid strain *i.e.* U-3. About 12 species of *Pleurotus* are also available in India for cultivation in different agro-climatic zones of India. By the continuous efforts of ICAR-DMR, 11 new strains were released for commercial cultivation *i.e.* DMR – Button 03, U3 – 54 (High yielding strain of white button mushroom). NBS-1 and NBS-5 (Browning resistant strains of white button mushroom). DMRO-240 and DMRO-484 (High yielding strain of Paddy straw mushroom). DMRO-38 and DMRO -388s (High yielding strain of Shiitake mushroom); DMR-334 (High yielding strain of milky mushroom); DMR-334 (High yielding strain of milky mushroom); DMR-*Macrocybe*-01 (High yielding strain of *Macrocybe* mushroom). The culture of all these strains are available from the culture bank maintained at the centre of ICAR-DMR-Solan. Presently Solan is the chief source of mushroom cultures in the country. They maintain cultures in their culture bank, refrigerator, liquid paraffin and in liquid nitrogen.

Spawn is prepared and supplied in 1 kg polypropylene packs and one kg of mushroom spawn costs about 70.00 Rs. per kg. Some big spawn companies are supplying spawn to commercial growers in India at almost 3 times the price of the available locally. The quality of spawn produced by Indian companies is comparable to the best in the world except that it is in small packs of 1 kg and the choice of strains is limited. India offers the lower cost of production of spawn with low cost of inputs. Presently Indian market demand is about 8000-10000 tons of commercial spawn. Out of this ICAR-DMR supplies 10.0 per cent of it. In last five years ICAR-DMR, Solan has supplied a total number of 2129 cultures of different mushrooms to different spawn production units, research labs and mushroom growers to prepare the quality spawn. Total amount of 52.1 tons of commercial spawn was sold to farmers and mushroom growers from the centre during the period 2012-2017. Private spawn growers and suppliers are fulfilling the demand of spawn. Higher production of spawn have been recorded from Punjab, Haryana, Uttarakhand, Tamil Nadu, Maharashtra and Odisha, where more number of private spawn labs are located.

Export Avenues

World mushroom production has gradually increased from 0.30 million tons to 3.41 million tons over a period of last 50 years from 1965 to 2015 (Singh *et al.,* 2017). Mushroom export/import has continuously increased during this period. On the export front till 1993, Indian contribution to the world trade was almost negligible. but it was heartening to note that for first time during 1994 India not only figured in the US imports but emerged as the second largest exporter canned mushroom by replacing Taiwan. During 2016-17, the total export of button mushroom from India is canned and frozen form stood at 1054 quintals and generated an income of Rs.7282.26 labs. Mushroom industry is getting support from Public and Private funding agencies and making rapid rise under the Govt. policy of liberalization and globalization.

India exporting mushroom mainly to USA, European countries and Hong-Kong. India exports mushroom in fresh form or processed form, 95 per cent export is of button mushroom. Mushroom spawn is imported from Netherland, France and Taiwan. Main competitor of India are Poland and China.

Q.1 What is the status of India in Export Industry ?

Marketing of Mushrooms

Mushroom cultivation is a reliable and effective way of resource for poor cultivars that they can grow nutritious food in limited space. It provides them opportunity to generate income by involving in highly tradable commodity. As shelf life of mushroom is maximum one week to so that producer have to face many problems. Due to poor marketing strategy farmers are not able to make proper profit from it (Shirur *et al.,* 2015). Marketing system is not organised, producer directly selling to retailer or consumers which has its own limitations, seasonal growing of mushroom also increases market problem. Here are some approaches which can make market successful:

- Adding value and increasing the shelf life of the mushrooms by creating processed products including dried or pickeled mushrooms, sauces, teas, cookies, bread, papad and pizzas can be prepared.
- Becoming organized and teaming up with other producers, to bulk up on volume and the variety of mushrooms and attract traders regularly to enable reliable sales of the perishable produce.
- Exploring various market options for fresh mushrooms – depending on transport infrastructure, selling directly to local customers, local traders, markets, intermediates, regional wholesalers, local Restaurants, shops or farmers co-operatives.
- Identifying existing markets and trading routes and identifying any niches to be filled (Sharma *et al.,* 2017).

Opportunities of Mushroom Production in India

India is not major producer of mushroom of any of varieties. But due to diverse climatic conditions it produces edible and medicinal varieties all over the India. Technical and non-technical manpower needed for the cultivation is available in sufficient amount.

Mushroom cultivation is based on recycling of agrowastes which are available in huge amount in the every state of the country. Carbon emission and black soots are another problem which is due to the burning of agricultural waste; states which are growing wheat and rice crops and facing great problems to deal with residual wastes. Mushroom cultivation can effectively utilize these agro-residues for agro production of protein rich food and plays crucial role in the management of these agro-residues.

The supply and demand gap in the World trade of mushrooms and the shrinkage of production in western countries due to high labour costs would result in better market prices for Indian mushroom producer. Due to its large population, India in itself can be a large market for its production. Development of infrastructure facilities and well organised distribution network provides the greater scope for marketing of perishable products in to meet domestic consumer demands. To achieve high goal in domestic and export market it is essential to produce quality fresh mushrooms and mushroom fortified value added products at competitive rates without any agro-chemical residues. Effort should also be made to exploit the commercial utilization of mushroom substrate left after cultivation for preparation of organic manure, vermi-compost *etc.*

Recently unemployment is increasing rapidly both in developed and developing countries. In this situation self employment can be an important way to increase employment. Mushroom cultivation not only provides a gainful employment to the rural Indian youths, but cost of mushroom production per unit area can be reduced. Government and cultivars keeping in view the increasing demand of mushroom due to globalization can compete to the international market.

Government Policies Regarding the Promotion of Mushroom Cultivation

Government policies are for the support of young entrepreneurs and farmers at state level or central level.

1. HMNH (Horticulture Mission for North East and Hilly States)

This scheme is of Government of India for promoting mushroom cultivation in Himalayan region under sub scheme of Integrated Development of Horticulture. Government is providing subsidy and financial assistance in Public and private sector for establishing the production unit, spawn production and compost making unit.

For Establishing

Production unit: 100 per cent financial assistance can be given upto the cost of 20 lakh/unit in public sector and 40 per cent in private sector.

Compost making unit: 100 per cent financial assistance can be given upto the cost of 20 lakh/unit in public sector and 40 per cent in private sector.

Spawn making unit: 100 per cent financial assistance can be given upto the cost of 15 lakh/unit in public sector and 40 per cent in private sector.

Some other schemes helps in Post Harvest management *i.e.*, Pre cooling unit, Cold storage and Marketing Infrastructure.

2. MOFPI (Ministry of Food Processing Industries)

This scheme provides grants up to 33 per cent of the cost of Unit and machinery in the hilly areas. One can avail help in up gradation and in establishment of Food Processing Industries. Under the scheme of Mega Food Parks, facilities has been provided for proper Infrastructure development of food processing unit along the value chain from the farm to market.

3. APEDA (Agricultural and Processed Food Products Export Development Authority)

This organization helps in the development of Export in Agriculture and Food Industry. Various schemes has been launched by APEDA for providing assistance to beneficiaries, R&D Organizations, Infrastructure Development,Transport Unit, Exporters, Producers and Quality Development and Maintenance.

4. NHB (National Horticulture Board)

The schemes under this Boards are for the establishment of Commercial Unit,

Primary processing and Post harvest management. The projects gets 50 per cent assistance as a subsidy in Hilly areas up to the cost of 37.50 Rs Lakh.

This board also helps in Post Harvest Management, relating to Pack House, Refrigerated Van, Retail Outlets, 50 per cent subsidy is given to the project of the cost of 72.50 Rs lakhs in hilly areas.

5. DMI (Directorate of Marketing and Intelligence)

This scheme is with the idea of Integrated Development of Marketing related to Agriculture and allied produce in the country. Purpose of this scheme is to strengthening and development of Agriculture marketing, Infrastructure, grading and develop infrastructure for post harvest requirement and marketing and management of surplus produce.

It may cover functional infrastructure like collection, drying, cleaning, grading standardization, sanitary and photo- sanitary measures, labelling, packaging,value addition and certification.

Assistance is also provided to post harvest operations like refrigerated vans used for transporting agricultural produce to maintain cold supply chain.

6. SFAC (Small Farmer's Agriculture Consortium)

It is a registered society comes under the Department of Agriculture and Cooperation, Government of India. It works for Agribusiness Development through venture capital assistance and project Development facility. Nationalised Banks participates in the promotion of Investment in Agribusiness projects. The venture capital is up to 10 per cent of total project cost of 75 lakh Rs in Agri business. Higher venture capital can be provided to farmers under special cases.

7. Rashtriya Krishi Vikas Yojna

Under this scheme Government provides assistance of 80,000 Rs. to mushroom growers for installation of racks, kits,tools, compost and spawn.

Initiatives taken by State Government for mushroom development.

a) Training on mushroom cultivation, 7 to 10 days training camps providing information regarding cultivation of mushrooms.

b) Preference to women, SC/ST and unemployed graduates in training.

c) It provides extension services to the mushroom growers.

d) It provides consultancy to the private sector and cooperative for setting of mushroom unit.

e) Helps in setting the marketing channel for the industry and suitable marketing.

f) Department of Horticulture provides good quality spawn, compost and casing soil to the mushroom growers.

g) NABARD and NHB provides bank loans to the mushroom growers.

Extension Service

a) Provides free of cost guidance to mushroom growers.

b) Distributes literature related to mushroom cultivation to the growers, whether they are small scale or large scale growers.

c) Make aware related to develop disease free compost and spawn.

QUESTIONS

1. **How Entrepreneurship can be established in India in field of mushroom cultivation?**
2. **Mention about various centres working in the field of spawn production.**
3. **Mention Govt. bodies working in the field of mushroom culture 2 spawn formation.**
4. **Discuss about the Govt. policies regarding the promotion of mushrooms cultivation.**
5. **Expand the following**

NHB	**DMI**
DMR-Solan	**SFAC**
RKVY	**HMNH**
MOFPI	**APEDA**

6. **Discuss about Export Avenues of mushroom in India?**

Chapter 11

World Scenario of Mushrooms

Mushrooms are edible fungi, now they have been classified as vegetables by some groups, debate is going on that they are part of horticulture or not. Technically they are not considered as plants *i.e.* due to absence of chlorophyll in mushrooms. Owing to their heterotrophic nature just like animals, fungi have different clade. They are rich in Potassium, Vitamins particularly Vitamin D, Selenium, Riboflavin and other ingredients that are beneficial for human health. Studies on edible mushrooms suggest that their consumption supports health, enhances immunity, weight management, and boost overall health. Consumption of mushrooms potentially reduces risk of diseases, such as breast and prostrate cancer. Mushrooms are gaining gradual recognition and is becoming crucial part of the diet not only in developed Countries but also in developing Countries. The commercial production and distribution of mushroom have improved significantly, by the expansion of the global mushroom industry. In forth coming years the growth of mushroom market is expected to increase because fast spreading awareness to mushrooms. The Global mushroom market size was valued at USD 50.3 billion in 2021 and it is expected that compound annual growth rate (CAGR) will expand of 9.7 per cent from 2022 to 2030. As the population of vegan food consumption is increasing day to day, demand of protein rich diet around the globe is expected to be a key driver for the market over the forecast period.

1. **Is mushroom a vegetable?**

2. **What are the health benefits of Mushrooms?**

Health Benefits

High medicinal value of the mushrooms and increased awareness about it among people. It is expected to contribute to the growth of the mushroom market within coming years. Growing demand of food that is low in carbohydrates fat content and high in nutrients such as Potassium, Selenium, Vitamins and other micronutrients is expected to escalate. The growth in global mushroom market is because of increased demand of mushrooms across Hotels, Cafeterias and Restaurants. It is predicted that, increasing adoption of mushrooms as a substitute for meat, coupled with increasing number of vegan populations, will further drive the growth of mushrooms.

The Governments across several countries are investing in the production and marketing of high-quality prized mushrooms that are not only safe for human consumption but are exported to other countries to generate revenue.

1. **Is there any policies of Government to promote Mushroom production ?**

Global Scenario

Mushroom industry has expanded horizontally and vertically globally, it spreading its wings in production and addition to newer types of mushrooms for commercial cultivation, both edible and non edible mushrooms. Today China is leading in world mushroom production of non edible and edible types. Mushrooms are China's sixth economically important crop as far as country's revenue generation is concerned. China produces about 70 per cent of world mushroom production. USA is the second highest mushroom producing country, followed by some European countries. France, Germany, Holland, Italy are the main mushroom producing countries in Europe. In Eastern European countries like Hungary and Poland has received a boost in production of mushrooms.

Global Mushroom Market

A recent survey shows that mushroom market had a value of 35 billion dollars in 2015. But it is expected to grow by 9.2 per cent between 2016 and 2021 and this would bring size to nearly 60 billion dollars in further years.

The biggest growth is expected in Asian countries. China remained leading global producer of mushrooms of all types. The Chinese National Government increasingly encouraged to shift their agricultural production out of traditional crops to high value crops like mushrooms and its processing. Mushrooms are accounting for more than 35 per cent of global market. Moreover demand is on the rise in North America and South America.

1. **Which country is the leading producer of Mushroom?**
2. **How many types of mushrooms are commercially grown ?**

UK market is good for mushrooms and sales are going well. UK is the largest market for polish exporters, due to its high consumption. Demand is high all year around, but it peaks in September and November, just before the holidays. Most of the mushrooms supplied to the British market are white and chestnut mushrooms but range has expanded now to shiitake, enoki and shimeji mushrooms. The United States and EU countries are the Second and third largest global producers (Karthick and Hamsalakshmi, 2017).

1. Is demand of mushrooms is increasing Globally?

Global Consumption and Trade

Mushroom consumption in China, Japan and India is quite large, due to growing focus on healthy and organic foods.

In Germany also, market is shifting from wild mushrooms to cultivated varieties. Demand of mushroom peaks towards Christmas. Besides white mushrooms there is also demand for exotic varieties such as Oyster, Shiitake, and Shimeji. Total volume of production has increased to 70,000 tones, which is increase of 3,000 tones compared to 2015.

Due to fall in the value of British currency, there is a lot of pressure on the market but that does not hit mushroom market. Special mushrooms are recording some growth but the demand for these products is still small. The growth is due to the increasing attention given to these mushrooms in media. Consumers want to try new colours and textures. Some of these mushrooms are wild, other such as Shiitake and Oyster mushrooms are grown in farms. Demand of mushrooms increases towards winter especially at the time of Christmas. About 7 tones of mushrooms are delivered every week to supermarkets and wholesalers in Edinburgh and London. Shiitake is in great demand in UK, barely they are able to keep up with demand.

Boletus edulis is a delicacy in France. Prices of *Boletus* are about 20 Euro per kilo, ten times higher than white mushrooms which cost about 2-3 Euro per kg. In Italy supply of cultivated mushroom is stable all year around, peak consumption takes place in autumn and winter. White mushroom are the most commonly consumed mushrooms in Italy. The cost is between 1.8 and 2.0 Euro per kilogram. Oyster mushroom follows in popularly with a price of 2.5 and 3.0 Euro per kg. Third and fourth are the *Pioppino* and Oyster which cost 4-9 Euro per kg. During autumn *Champignon* and *Cardoncello* mushrooms are grown in Sardinia. The prices in the whole sale market of Rome and Turin range from 1.70 to 1.85 Euro of white mushrooms to the higher prices of Exotic mushrooms. In Verona, the *Chatharellus* cost 9.14 Euro per kilo, the oyster mushroom from Veneto stand at 3.20 Euro per kg, while cost 10 Euro per kg and Porcino reaches 25.50 Euro per kg.

US is mainly dependent on wild mushrooms. They are picked in Montane, California, Oregon and Washington. The season normally starts in August and lasts until October. Market mainly depends upon the natural phenomena. The prices are high due to import duty of 22 per cent on wild mushrooms. Imports arrive from Eastern Europe and South Africa among other places. *Agaricus* mushrooms are the most cultivated in the US. In 2015/2016 their production was 461,000 tones.

They also produce brown varieties like *Portobello* and *Crimini*. Between 2013 and 2016, market for brown mushroom grew by 8.3 per cent, while in the same period *Agaricus* category recorded a 40 per cent higher than the previous season.

With in the mushroom market as a whole, the share of organic mushrooms is growing. The total production is 2015/2016 campaign reached 45,550 tones an increase of 49 per cent compared to previous season. *Agaricus* account for 2/3rd of the organic market. In total, organic mushrooms have a share of 7 per cent of the market.

Brazilian market is growing about 20 per cent per year. Most of the market demand is covered with domestic production, although there are small imports from China. But producers of Oyster and Shiitake want to grow along with the market.

Domestic market in South Korea is stable but the producers want to be bigger they are looking for new market overseas. They are growing wide varieties like oysters, enoki and Shimeji mushrooms. These mushrooms are exported to Netherland, Germany, Switzerland, Spain, Greece, Austria, Poland, Indonesia, U.S, Vietnam and Australia. As these mushrooms are perishable so their global movements often have been restricted to transactions mainly between neighboring countries.

According to F&B Report (2019), mushroom is growing commercially in almost all the major mushroom growing countries such as China, USA, Italy, UK, Germany, Taiwan and others. The demand for Shiitake mushroom is expected to grow in the upcoming years as the consumption of Shiitake in South Asian Countries and European countries is increasing at a decent pace. Demand of fresh mushroom is increasing across the globe, U.S. and other developed economies, as the demand of healthier food products is increasing in which preservatives are not added. The U.S. is one of the largest producers of fresh mushrooms. The consumption of mushroom is maximum in China. The consumption of mushroom in Asian countries such as Japan, India and others is increasing at a significant pace attributed by increasing production along with the increased import of mushrooms in these countries. Adoption of advanced and modern mushroom cultivation techniques by mushroom farmers of the region is expected to fuel the mushroom market of the region in the coming years.

The U.S was the second largest producer accounting for approximately 375 million kg in the year 2019. As mushrooms is in the list of superfood, consumption is increasing. The U.S Government is continuously increasing import duties, which is also contributing to the higher prices of mushrooms.

Present Status in India

Mushroom as first cultivated in India in 1940, however its systematic cultivation was first attempted in 1943, Department of Agriculture, Solan, Himachal Pradesh, started the work on a small scale to grow mushrooms. In India "Guchhi" mushroom was a delicacy known to people prior to 1950's. It came mostly from Kashmir. In

India present production of mushroom is around 30,000 M.T. as per the information available from APEDA (2001-2002) 15897 M.T. of mushroom of the value of Rs. 72,47,54,829 is exported. Price of morel mushroom is about 38,000/Rs. per Kg. The Global mushroom market was valued at 12.74 M.T. in 2018 and to reach 20.84 M.T. by 2026. Around 1950's the Government of Himachal Pradesh appointed Dr S.S. Jain, first Asstt. Plant Pathologist and Mycologist for the State. He started the Research Centre in Solan for growing edible mushroom *Agaricus* and other species. Now the mushroom growing in India has increased many folds. Mushroom cultivation has become a huge export oriented Industry and large foreign exchange earning business and also profitable for small time growers; Many Universities and State Departments of Agriculture as also private people are giving training in growing mushrooms, which are mostly exported and also used in many good hotels in a variety of culinary delights.

In early Seventies, mushroom cultivation was started spreading to other hilly regions of U.P. and Tamil Nadu. By early Eighties mushroom cultivation was adopted by farmers in Haryana, Punjab, UP and in areas around Delhi, as a seasonal crop. Delhi and Bombay became big market together for fresh mushroom and utilizes more than 10 to 12 tons of fresh mushrooms per day. Now mushroom cultivation is also picking up in the Punjab, Haryana, Rajasthan, M.P., Maharashtra, Karnataka, A.P. and other states. *Pleurotus* was mainly growing in North eastern states due to its weather conditions.

Industry Key Players

According to F and B (2019) report the mushroom market in many developed and developing economies is proliferating at a significant pace. Their distribution is a challenge across many countries. Investment by industrialist as well as Government to improve the distribution system of fresh and processed mushroom can bring immense growth opportunities for the sales of mushroom across the world. Numerous industrialists are focusing on merger and acquisition strategies to remain prominent in the global market.

List of Key Companies

- Biofungi GmbH
- Agro Dutch Industries Ltd.
- Weikfield Foods Pvt. Ltd.
- California Mushroom Farms Inc.
- Highline Mushroom
- Monterey Mushroom Inc.
- Nasza Chata
- Banken Champignous B.V.
- Bonduelle
- Ecolink Baltic.

The increasing demand for mushroom which is leading to increasing investments by industrialist and government to set up infrastructure for cultivation of edible mushroom is a significant growth engine of the global mushroom market share. By growing and doing business properly. One will get good profits.

In Conclusion, one can say that Mushroom business is one of the most profitable agribusiness that you can start with low investment and less space. Mushroom cultivation is India is growing gradually as an alternative source of income for many people. One can even export mushrooms to other counties. Mushroom business has huge potential in India because climatic condition for mushroom farming business is great in India as well big market for consumption.

QUESTIONS

1. **In India what is the status of mushroom farming?**

2 **Which are the main mushroom growing states in India?**

3. **Name some mushroom growing companies in India and in World.**

Bibliography

Annual Report 2016-2017. All India Coordinated Research Project on Mushroom, ICAR – Directorate of Mushroom Research, Kanal, S., Attri, B.L. (ed). Chambaghat, Solan –pps 173-213 HP.

Annual Report 2017-2018. Department of Agriculture, Cooperation and Farmers Welfare, Ministry of Agriculture and Farmers Welfare, Government of India Krishi Bhawan, New Delhi-110001.

Atkins, P. and Akino, 1971. Major diseases of the cultivated white mushrooms. MGA. 1-23.

Baker, (1934) Effect of different cultivation techniques on the field of *Volvariella*. https://shodhganga.inflibnet.ac.in.

Bokaria, K., Balsundram, S.K. Bhattaria, I. and Kaphle, K. (2014). Commercial Production of Milky mushroom (*Calocybe indica*). Merit Research Journal of Agricultural Science and Soil Sciences (ISSN: 2350-2274) vol. 2 (2) pp. 32-37.

Carrera, D.M. (1998) Cultivation of Oyster Mushroom. McGraw Hill Year Book of Science and Technology. 1999. Ed: M.D. Licker pp. 242-245. ISBN 0-07-0526225-7(447pp) (http://books.mcgraw-hill.com).

Chandravadana, M.V., Vekateshwarlu, G. Babu, C.S., Roy, T.K., Shivashankara, K.S., Pandey, M., Tewari, R.P. and Setvaraj, Y. (2005). Volatitle flavour components of dry milky mushroom. Flavour fragr J. 20: 715-717. (Google Scholar).

Chang, S.T. (1969) Factors affecting spore germination of *Volvariella volvacea*. https://online library. Wiley, con. Abs.

Chang, S.T. (1971). *Volvariella volvacea* and its life history. Jstor https://ww.jstor.org> stable.

Chang, S.T. (1999). World production of cultivated and medicinal mushrooms in 1997 with emphasis on *Lentinus ebodes* (Berk.) Sing, China. Int. J. Med. Mush. 1: 291-300.

Chen, A.W. (1999). Cultivation of the medicinal mushroom *Ganoderma lucidum* (Cart: Fr) P. Karst. (Reishi) in North America. International Journal of Medicinal Mushroom 1:263-282.

Chen, A.W. (2002). Natural log cultivation of the medicinal mushroom, *Ganoderma lucidum* (Reishi). Mushrooms Grower's Newsletter 3: 2-6.

Chen, A.W. (2014). Part-III. Mushroom for the tropics. Growing *Ganoderma mushrooms.* Mushroom Grower's Handbook 1: 224-234.

Chen, T.Q., Li, K.B., He. X.J., Zhu, P.G., Xu, J. (1998). Micro-morphology, chemical components and identification of log-cultivated *Ganoderma lucidum* spore. J. Proc. 98 Nanzing Intl. Symp. Science and Cultivation of mushrooms. 214.

Chyiu, S.W., Wang, Z.M., Leung, T.M. and Moore, D. (2000). Nutritional value of *Ganoderma* extract and assessment of its genotoxicity and antigenotixicity using comet assay of mouse lymphocytes. Food Chem Toxicol. 38: 173-178.

Conte, Anna. Del. and Laessoe, Thomas (2008). The edible Mushroom Book Dorling Kindersley p. 91. ISBN 978-1-4053-3213-2.

DGCI and S. Director General of Commercial Intelligence and Statistics Report. Ministry of Commerce and Industry, Government of India.

Directorate of Mushroom Research (ICAR). Salon India (2011). Mushroom cultivation, marketing and consumption. Directorate of Mushroom Research (Indian Council of Agricultural Research).

Food and Agriculture Organization of the United Nations (FAO). Available online http://faostat.FAO.org/site/567/Desktop Default.aspx?

Francia, C., Rapior, S., Courtecerisse, R. and Siroux, X. (1999). Current research findings on the effect of selected mushrooms on cardiovascular diseases (https://web.archieve.org/web/20110718164609).

Geofrey, Kibby (2003). Mushrooms and toad stool of Britain and Northern Emope. Hamlyn. P. 225. ISBN 978-0-7537-1865-0.

Habiyamnic, J. and Berovic, M. (2000). The relevance of solid – state subtrates moisturing on *Ganoderma lucidum* biomass cultivation. Food Technology and Biotechnology 38: 225-228

http://krishikosh. Egranth, ac. In/handle/1/581018341

International Journal of Medicinal Mushrooms. 1: 169-172.

Irawati, D., Hayashi C., Takashima, Y., Wedatama, S., Ishiguri, f., Yoshizawa, N., and Yokota, S. (2012). Cultivation of the edible mushroom *Auricularia polytricha* using sawdust – based substrate made of three Indonesia Commercial plantation species, Falcatarie, *Shorea* sp and *Tectona grandis*. Micologie applicade international, 24(2), pp. 33-41.

Jagdeesh, R., Lee, S.K., IM, J.H., Oh. M.J., Oh, X.L. and Jang, K.Y. (2018) Current prospects of mushrooms producton and industrial growth in India. Journal of Mushroom. 239-249. https://www. researchgate. net/publication/326682608.

Jitender, G. and Vaidya (2000). Traditional Medicinal Mushrooms and Fungi of India. Int. J. Med. Mushrooms. 2: 209-214.

Jonathan, S.G. Bawo, D.D., Adejoye, D.O. and Briyai, O.F. (2009.) Studies on biomass production in *Auricularia polytricha* collected from Wilber-force Island, Bayelsa state, Nigeria, AM. Appl. Sci. 6:182-186.

Kaliyaperamal, M. (2013). Molecular taxonomy of *Ganoderma cupreum* from Southern India inferred from ITS rDNA sequence analysis. Mycobiology 41: 248-251.

Karthick, K. and Hamsalakshmi, (2014). Current Scenario of Mushroom Industry in India. International Journal of Commerce and Management Research Vol. 3: Issue 3; pp.23-26. ISSN: 2455-1627. www.managejournal.com

Karthick, K., Hamsalakshmi (2017). Current Scenario of Mushroom Industry in India. India. J. Commerce Manag. Res. 3: 23-26.

Kiran Kumar, N., Krisnamoorthy, A.S., Kanalakannan, A. and Amirthan, D. (2016). Influence of temperature and pH on mycelial growth and Chlamydospore production of Paddy straw mushroom *Volvariella volvaceae*. J. Res. Angrau 44 (1 and 2) 1-7.

Krishnamoorthy, A.S. and Balan V. (2015). A comprehensive Review of Tropical Milky White Mushroom (*Calocybe india* P and C) Mycobiology: 43(3): 184-194. PMIP: 26539033.

Li, F., Wen, Ha, Lium, X., Zhou, F. and Chen, G,. (2012). Gene cloning and recombinant expression of a novel fungal immnomodulatory protein from *Trametes versicolar*. Protein Expression and Purification 82: 339-344.

Lin, S.B., Li, C.H., and Lee, S.S. (2003). Triterpene enriched extracts from *Ganoderma lucidum* inhibit growth of hepatoma cells via suppressing protein kinase C, activating mitogen-activated protein kinases. Life Science – Elsevier.

Man, J.L., Lin, H.C. and Chen, C.C. (2001). Non volatile components of several medicinal mushrooms. Food Res. Int. 34: 521-526.

Manjit, S., Kamal S. and Sharma, V.P. (2017). Status and trends in World Mushroom Production. J.. Mushroom Research, 26: 1-20.

Mehta, B.K., Jain, S.K. Sharma, G.P., Dashi, A. and Jain, H.C. (2011). Cultivation of button mushroom and its processing: A techno-economic feasibility. International Journal of Advanced Biotechnology and Research 2(12): 201-207.

Munjal, R.L. and Seth, P.K. 1974. Brown plaster mould – a new disease in white button mushrom. Indian Host. 18(4): 13.

Mushrom market size, share and industry analysis, By type, Form and Regional Forecast 2019-2026. Region: Global – Published Date Oct., 2019., Report ID: FBI100197. Food and Beverages – Mushrom Market.

Mushroom Cultivation: Cultivation and consumption of mushroom in Punjab. (.//././post agriculture – classifieds, html).

Mushroom, http://usda.mannlib.cornell.edu/usda/current/mush/mush-08-21-2017.pdf, August 21, 2017, by the National Agricultural Statistics Service (NASS), Agricultural Statistics Board, United States Department of Agriculture (USDA) ISSN: 1949-1530, Accessed on April 5, 2018.

Nilson, S. and Persson, O. (1977) Fungi of Northern Europe 2; Gill – fungi, Penguin, p. 24. ISBN. 014-063006-6.

Ohga, S. (2000). Influence of wood species on the saw dust based cultivation of *Pleurotus abalonus* and *P. eryngii*. J. wood. Sci. 46: 175-179.

Pandey, M. (2005). Mushroom Research and Development of India. http://www. mushworld. Com/Overases. Asp?_id=514- and vid = 7165.

Pani, B.K. (2011). Effect of age and quantity of spawn on milky mushroom production. Asian J. Exp. Biol. Sci. 2: 769-771. (Google Scholar).

Paterson, R.R. (2006). *Ganoderma* a therapeutic fungal biofactory. Phytochemistry. 67(18): 985-2001.

Pavel, K. (2009). Chemical composition and nutritional value of European species of wild growing mushroom: A review. Food Chemistry 113(1): 9-16.

Philips Roger (2006). Mushrooms, Pub-Mcmilan, ISBN 0-330-44237-6. P. 266.

Priya, R.U., Geetha, D. and Darshan, S. (2016). Biology and cultivation of Black ear Mushroom *Auricularia* sp. Advances in life Sciences 5 (22): ISSN 2278-3849, 10252-10254, 2016.

Purkayastha, R.P. and Chandra, A. (1974). New species of edible mushroom from India. Trans. Br. Mycol. Sco. 1974: 62: 115-118.

Rajapakse, P. (2011). New Cultivaton Technology for Paddy Straw mushroom (*Volvariella volvacea*). Proceedings of the 7th International Conference on mushroom Biology and Mushroom products (ICMBMP7) pp 446-451.

Raman, Jagadeesh, Lee, S.K., M.H., Oh, M.J., Oh, Y. Lee and Jang, K.Y. (2018). Current prospects of mushroom production and industrial growth in India. https://www.researchgate.net/publication/326682608

Rapior, S., Marrion, C., Pelissier, Y. and Bersiere, J.M. (1997). Volatite composition of fourteen species of fresh wild mushrooms. J. Essent Oil Res. 9; (231-234). (google scholar)

Reyes, R.G. (2000). Indoor cultivation of Paddy straw mushroom, *Volvariella volvacea* in crates. Mycologist, Vol. 14 (4): 174-180.

Rohan, Mushroom market worth $ 50,034.12 million by 2019. Markets and Markets, https://www.marketstandmarkets.com/press relese/mushroom asp. Accessed on April 5, 2018.

Schrot, J. (1888). Mycobank. International Mycological Association. Retreived, 20 Sep. 2010. http://www.mycobank.org/mycotaxo aspx? Lin k. T_2 Rec. 102281).

Seth, P.K. and Dar, G.M. (1989). Studies on *Cladobotryum dendroides* (Boll. Merat) W. Games et Hoozen; causing cobweb disease of *Agaricus bisporus* Sing. Mushroom Science, 12: 711-723.

Seth, P.K., Kumar, S. and Shandilya, T.R. (1973). Combating dry bubble of mushrooms. Indian Host. 18(2): 17-18.

Sharma, S.K., Lall, A.M., Sharma, M. and Reishi, M. (2013). Response of Organic supplementation on yield and nutritional parameters of *Calocybe indica* vegetos. 26:36-39. (Google Scholar)

Sharma, S.R., Kumar, S. and Sharma, V.P. (2006). Physiological requirement for cultivation of Malaysian strain of Shiitake, *Lentinula edodes*. J. Mycol. Pl. Pathol. 36: 149-152.

Sharma, V.P., Annepu, S.K., Gautam, Y., Singh, M. and Shwet Kamal (2017). Status of mushroom production in India. Mushroom Research 26(2): 111-120.

Sharma, V.P., Kumar, S. and Annepu, S.K. (2017) Technologies developed by ICAR – Directorate of Mushroom Research for Commercial use. Technical Bulletin pp 11-13.

Shirur, M. and Shivalingegowda, N.S,. (2015). Mushroom marketing channels and consumer behaviour: A Critical Analysis, Mysore Journal of Agricultural Sciences 49(2): 390-393.

Shukla, S. and Jaitely, A.K. (2013) Effect of temperature on mycelial growth of different strain of *Calocybe indica* mushroom. Online International Journal of Biosolution 3 (1): 121-123.

Singer. R. 1961. Mushrooms and Truffles: Botany, Cultivation and utilization: 1-272. www.mycobank.org Biolomics.

Singh, Manjit, Shwet Kamal and Sharma, V.P. (2017). Status and trends in world production-I. Mushroom Research 26(1): 1-20.

Singh, R.P. and Mishra, K.K. (2005). Mushroom cultivation. Mushroom Research and Training Centre, G.B. Pant University of Agriculture and Technology Pant Nagar – 263145.

Singh, Y. and Sidhu, H.S. (2014). Management of Cereal Crop residues for sustainable rice-wheat production system in Indo-Gangetic plains of India. Proc. Indian Nat. Sci. Academy, 80: 95-114.

Sohi, H.S. and Kumar, S. (1986). Disease of edible mushrooms and their control NCMRT. Chambaghat, Solar, Ext. Bull. No.2: 12 pp.

Sohi, H.S., Seth, P.K. and Kumar, S. (1965). Occurrence of truffule disease (*Pseudobalsamia microspora* Delhi and Ilambert) in mushroom beds. Curr. Sci. 31: 488-489.

Stamets, P. (1993) Growing Gourmet and Medicinal mushrooms. Ten speed press. Berkeley. 56p.

Stamets, P. (2000) "The Role of Mushrooms in nature". Growing Gourmet and medicinal mushrooms Shokyo Oyobi Xakugo kinoko no sabai (3rd edn., Barkeley california, USA; ten speed pres pp 10-11. ISBN – 978-1-58008-175-7.

Sterry, Paul and Barry, Hughes, (2009). Complete Guide to British Mushrooms and Toad stools. Harper collins p. 290. ISBN 978-0-00-723224-6.

Stoller, B.B. (1943). The use of ammonia in controlling the truffle disease of mushroom. Phytopathology 33.

Tang, L; Xia, X; LiLi., Gno, Cr; and Bian, Y. (2010). Analysis of genetic diversity among Chinese *Auricularia auricula* cultivar using combined ISSR and SRAP markers. Current Microbiology 61(2):132-140. Doi.10.1007/500284-010-9587-4(http://doi.org/10.100752Fs00284-010-9587-4). PMID 20127246 (http://PubMed.ncbi.n/m.nih.gov/20127246).

Thakur, A., Rana, M., Lakhanpal, T.N., Ahmad, A. and Khan, M.I. (2007). Purification and characterization of lectin from fruiting body of *Ganoderma lucidum.* Lectin from *Ganoderma lucidum.* Biochem Biophys Acta 1770: 1404-1412.

Tiwari, R.P. and Singh, S.J. (1984). Mushroom virus disease in India. The mushroom Jour. 142: 354-355.

Tomblin, Gill (2007). How to identify Edible Mushrooms. Harper collins UK p. 146. ISBN 978-0007259618.

VanDer Hens, L., Van Dr Vilet, A., Bocka, C.F.M., Kino, K., Hoitsma, A.J., Tax, W.J.M. (1995). Lingzhi polyphosphoric fungus. In: Herbal and Traditional Medicine: Molecular aspects of Health. New York: Marcel Dekkar, pp: 179-228

Verma, R.N. (2002). Recent Advances in the cultivation technology of Edible Mushroom, NRCM solan 1-10.

Wakchaure, G.C. (2011). Production and marketing of Mushrooms; Global and National Scenario. Mushrooms Cultivation, marketing and consumption Global and National Scenario. Pp 11-22. Manjit Singh, Bhunesh vijay, Shewt Kamal, G.C. Wakchaure (eds). Directorate of Mushroom Research, Solan, Himachal Pradesh – 173213, India. https://www.researchgate.net/publication/235951347.

Wang, H., Ng, T.B. (2006). Ganodermin, an antifungal protein from fruiting bodies of the medicinal mushroom *Gonaderma lucidum* Peptides. 27: 27-30.

Wang, P.Y., Zhu, X.L. and Lin, Z.B. 2012. Antitumor and immunomodulatory effects of polysaccharides from broken spore of *Ganoderma lucidum* Frontiers in Pharmacology.

World Population Clock: 7.7. Billion People (2019) – Worldmeters. Into Archieved from the original on September 2016. Retrieved 27th April, 2019.

World Population Lits 6 billion. 4 March 2004. Archieved from the original on 22 Feb., 2018. Retrieved 11 Feb., 2018.

World population to feet milestone with Birth of 7 Billionth Person. PBS News Hour. 27 October 2011. Archived from the original on 24 Sept. 2017. Retrieved 11 Feb. 2018.

Wuest, P.J., Cole, H. and Sanderr P.L,. (1974). Tolerance of *Verticillium* malt to benomyl. Phytopathology, 64: 331-334.

Yoder, J.B., Sinden, J.W. and Houser, E. (1950). Experience with Zinc ethylene biodithiocarbamate as a fungicide in mushroom cultivation. Mushroom Science, 1: 100-108.

Yuan, Z., He, Puming., Cui, J., Takench, H. (1998). Hypoglycemic effect of water soluble Polysachharide. From *Auricularria auricula* Judae cruel on genetically diabetic KK0Ay mici (https://doi.org/10.1271/bbb.62.1898).

Zadrazil, F. (1992). Conversion of lignocellulose in to feed with the white rot fungi. In: Solid substrate cultivation, pp. 321 – 339 (eds) H.W. Doelle, D.A. Mitchell and C.E. Rolz. Elsevier Publication, Essex. England, pp. 321-339.

Zhou, X. and Deng, B. (1996). Research on various media affecting mycelial growth of *Granoderma*. Journal of Hanhang Teachers College – en.enki.com.ca.

Zhou, Y., Qu, Z., Zang, Y. and Lin, X. (2012). Neuroprotective effect of pre administration with *Ganoderma lucidium* spox on rat hippocampus. Experimental and Biological Research – Elsevier.

Index

COLOUR PLATES

Figure 1.5a: ***Agaricus*** (Button Mushroom) (pg.10)

Figure 1.5b: ***Lentinula*** (Shiitake) (pg.10)

Figure 1.5c: ***Pleurotus florida*** (White Oyster) (pg.11)

Figure 1.5d: ***Pleurotus djamor*** (Pink Oyster) (pg.11)

Figure 1.5e: ***Auricularia*** (Ear Jelly mushroom) (pg.12)

Figure 1.5f: ***Calocybe*** (Milky Mushroom) (pg.12)

Figure 1.5g: ***Portebello*** (Brown Mushroom) (pg.12)

Figure 1.5h: ***Volvariella*** (pg.13)

Figure 1.5i: ***P. columbinius*** (Grey Oyster) (pg.13)

Figure 2.22: Solar Dryer (pg.38)

Figure 3.1: ***Agaricus*** (Button Mushroom) (pg.42)

Figure 3.24: Bed of Button Mushroom (pg.58)

Figure 3.25: Bags of Button Mushroom (pg.58)

Figure 3.26: Fruiting of Button Mushroom (pg.59)

Figure 3.27: Harvesting of Mushroom (pg.59)

Figure 3.28:After harvesting of Mushroom (pg.61)

Figure 4.1: Fruiting bodies of ***Pleurotus.*** (pg.67)

Figure 4.2a: ***P. sajor-caju*** (pg.68)

Figure 4.2b: ***P. florida.*** (pg.68)

Figure 4.2c: ***P. columbinus*** (pg.68)

Figure 4.2d: ***P. djamor*** (pg.69)

Figure 4.2e: ***P. ostreatus*** (pg.69)

Figure 4.2f: ***P. flabellatus*** (pg.69)

Figure 4.2g: ***P. citrino-pileatus*** (pg.70)

Figure 4.9: Commercially available Spawn (pg.76)	Figure 4.10b: Placing of Bags in racks. (pg.77)	Figure 4.10c: Bags can be placed on floor. (pg.77)

Figure 4.10d: Fully grown mycelium in Bags. (pg.77)	Figure 4.10e: Fully grown mycelium in Bag (pg.78)	Figure 4.11(a & b): Initial pinning of fruiting bodies. (pg.79)
		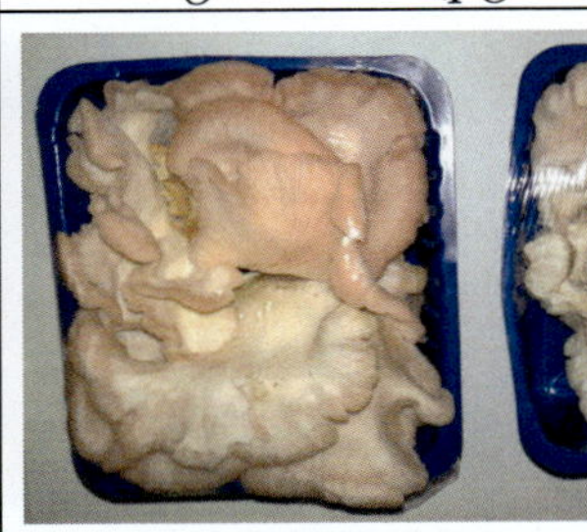
Figure 4.12: Various Stages of fruiting bodies. (pg.80)	Figure 4.13: Packaging (pg.80)	Figure 4.14: Packaging (pg.81)

Figure 5.1: Fruiting Body of ***Calocybe*** (Milky Mushroom). (pg.88)	Figure 5.5: Mycelium on the surface. (pg.92)	Figure 5.6: Bed preparation. (pg.92)

Figure 5.8a: Pinning of fruiting body on the compost. (pg.94)

Figure 5.8b Pinning stage (pg.94)

Figure 5.8c Young fruiting body (pg.95)

Figure 5.8d Mature fruiting body (pg.95)

Figure 5.9 Harvested fruiting body (pg.95)

Figure 6.1: Fruiting body of ***Ganoderma*** (pg.102)

Figure 6.2: Fruiting body on solid substrate (pg.105)

Figure 6.5: Mixing of substrate (pg.108)

Figure 6.6: Sterlization of PP Bags (pg.110)

Figure 6.7: Filling of PP Bags with substrate (pg.110)

Figure 6.8: Filled PP Bags with susbstrate (pg.110)

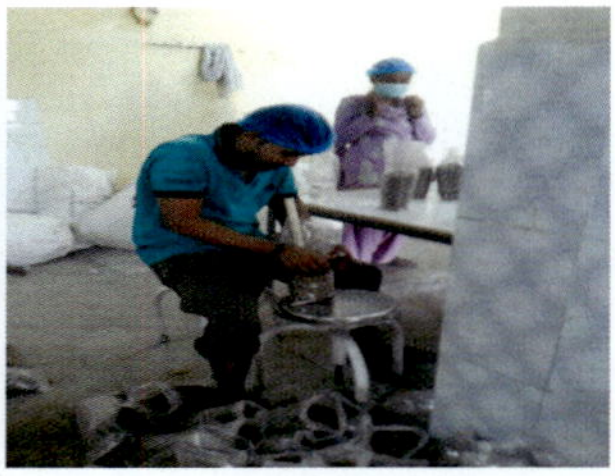

Figure 6.9: Making of hole in filled PP Bags (pg.111)

Figure 6.10: PP Bags with hole (pg.111)

Figure 6.11: PP Bags with substrate ready to autoclave (under process) (pg.111)

Figure 6.12: PP Bags with substrate ready to autoclave (pg.112)

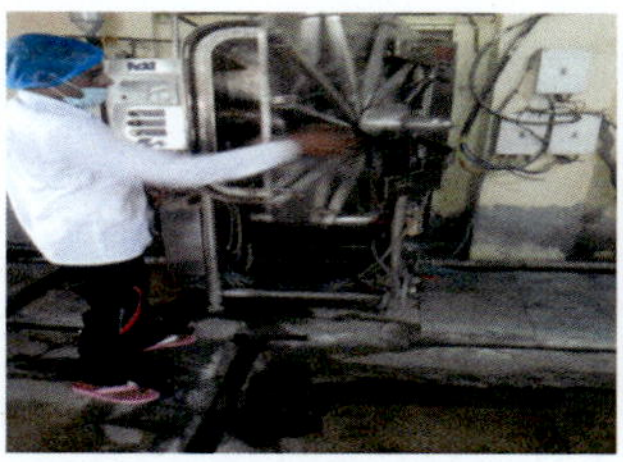
Figure 6.13: PP Bags with substrate in autoclave (pg.112)

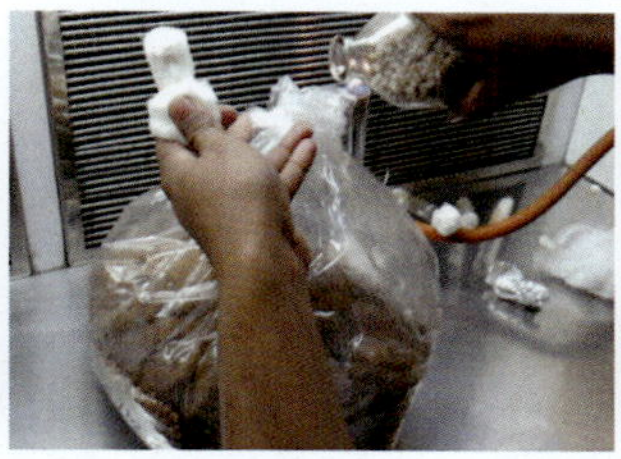
Figure 6.14: Inoculation of PP Bags with spawn (pg.112)

Figure 6.15: Incubation of PP Bags with substrate and spawn (pg.114)

Figure 6.16: Fruing of ***Ganoderma*** in Bags (pg.114)

Figure 6.17: Chips of ***Ganoderma*** (pg.116)

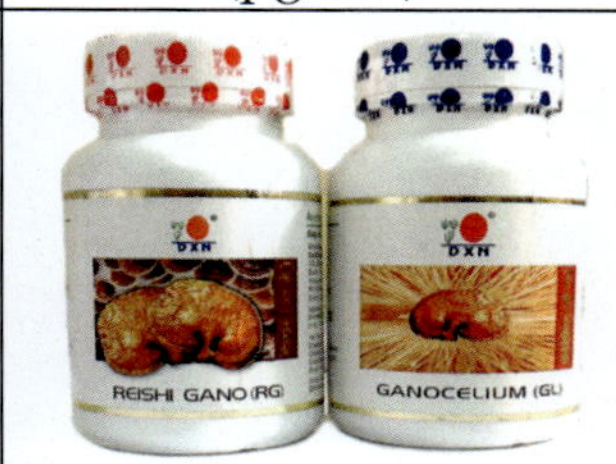

Figure 6.18: Commercially available tablets of ***Ganoderma*** (pg.117)

Figure 6.19: Mature Fruiting body of ***Ganoderma*** (pg.117)

Figure 6.20: ***Cordyceps*** (pg.120)

Figure 6.21: ***Hericium coralloides*** (pg.121)

Figure 6.22: Fruiting body of ***Inonotus obliquus*** (pg.122)

Figure 6.23: Fruiting body of ***Amanita phalloides*** (pg.124)

Figure 6.24: Fruiting body of ***Amanita muscaria*** (pg.125)

Figure 6.25: Fruiting body of ***Amanita virosa*** (pg.125)

Figure 6.26: Fruiting body of ***Coprinopsis atramentaria*** (pg.126)

Figure 6.27: Fruiting body of ***Cortinarius orellanus*** (pg.127)

Figure 6.28: Fruiting body of ***Psilocybe cubensis*** (pg.127)

Figure 7.1: Fruiting body of ***Volvariella*** (pg.130)

Figure 7.3: Different systems of cultivation (pg.133)

Figure 7.4: Bundle preparation (pg.134)

Figure 7.5: Bed preparation and spawning (pg.135)

Figure 7.6: Different stages of growth of ***Volvariella*** (pg.135)

Figure 7.7: Spawn preparation (pg.138)

Figure 7.8: Croping & Harvesting (pg.139)

Figure 7.9: Post harvest management of ***Volvariella*** (pg.140)

Figure: 8.1: Fruiting body ***Auricularia*** (pg.144)

Figure: 8.1: Fruiting body ***Auricularia*** (pg.145)

Figure: 8.3 ***Auricularia*** fruiting bodies growing on the substrate (pg.146)

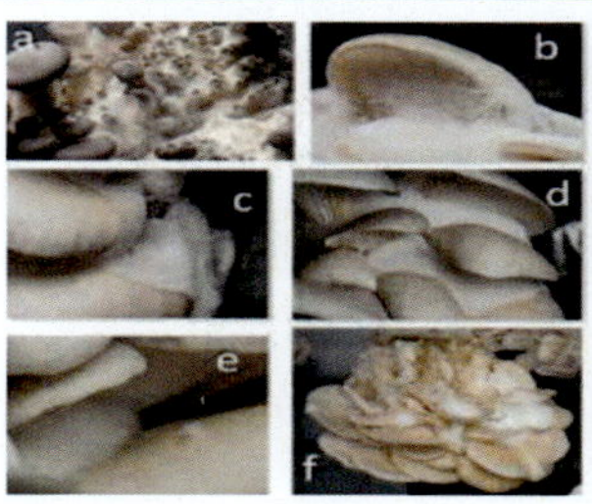

Figure 9.1: ***Dactylium cladobotryum*** on ***Pleurotus*** fruiting body. (pg.153)

Figure 9.2: Green Mould (***Trichoderma***) (pg.154)

Figure 9.3: Dry Bubble (***Verticillium fungicola***) (pg.155)

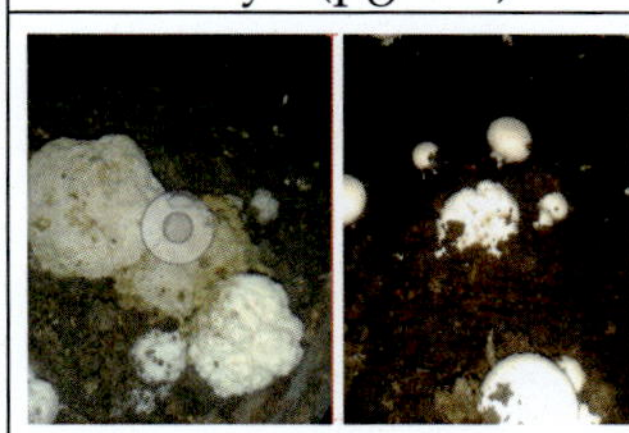

Figure 9.4: Wet Bubble (***Mycogone***) (pg.156)

Figure 9.5: Mildew disease (***Dactylium dendroides)*** (pg.157)

Figure 9.6: Brown Plaster (***Papulospora byssina***) (pg.157)

Figure 9.7: White Plaster ***(Scopulariopsis copulariopsis fimicola)*** (pg.158)

Figure 9.8: False Truffle (pg.159)

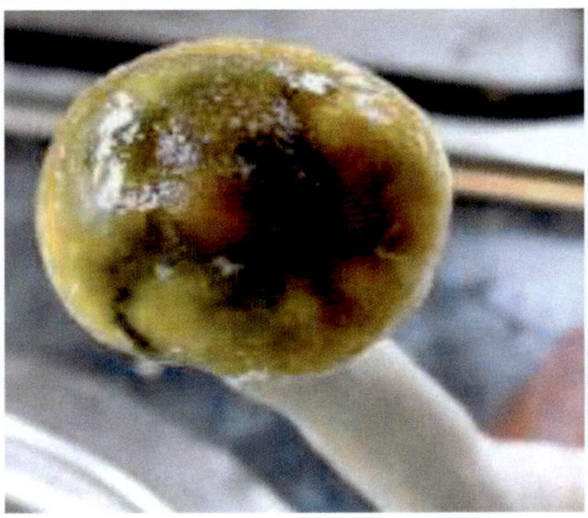

Fgure 9.9: Bacterial Blotch (***Pseudomonas fluorescens***) (pg.160)

Figure 9.10: Ink Caps (***Coprinus atramantarius***) (pg.161)